The R.A.M.S. Library of Alchemy

Volume 47

Selected Chemical Universal and Particular Processes

By

Alexius von Ruesenstein

R.A.M.S. Publishing Company

Selected Chemical Universal and Particular Processes

By

Alexius Baron von Ruesenstein
Leipzig 1664

Published in Frankfurt and Leipzig, 1754

Translated by Christine Bannerji

Produced by

Restorers of Alchemical Manuscripts Society

R.A.M.S. Publishing Company

R.A.M.S. Publishing Company
117 Rutherford Lane
Stuarts Draft VA 24477

The R.A.M.S. Library of Alchemy, Volume 47:
Selected Chemical Universal and Particular Processes
Copyright © 2016 R.A.M.S. Publishing Company

First Edition 2016

ISBN-13 **978-1523799664**
ISBN-10 **1523799668**

Image Processing by Philip N. Wheeler

Printed in the United States of America

Table of Contents

Dedicated to Hans W. Nintzel,

American Alchemist

and

Founder of the

Restorers of Alchemical Manuscripts Society

(R.A.M.S.)

Disclaimer

Liability: The publisher does not warrant or assume any legal liability or responsibility for the accuracy, completeness, or usefulness of any information, apparatus, product, or process disclosed. The publisher makes no representation as to the accuracy or completeness of the contents of this book and specifically disclaims any implied warranty of merchantability or fitness for a particular purpose. No warranty may be created or extended by written sales materials or sales representatives. You should obtain professional consultation where appropriate. The publisher shall not be liable for any loss of profit or other commercial or personal damages, including but not limited to special, incidental, consequential, or other damages.

Introduction

Philip N. Wheeler

This work was added to the R.A.M.S. Library in 1988 by Hans W. Nintzel. Here is the full title:

Selected Chemical

Universal and Particular Processes,

Which Baron von Ruesenstein

Learned on his Two Journeys with Six Adepts:

Gualdo, Schulz, Fauermann, Koller, Fornegg and

Monte Schider,

Many of which he tried himself and which he collected together

Personally in the year 1664, the originals of which were

Found in his castle in a wall.

Frankfurt and Leipzig, (Peter Conrad Monath) 1754.

And now translated from the German tongue into King's English

For the lovers of the Philosophical Labours;

September 1988.

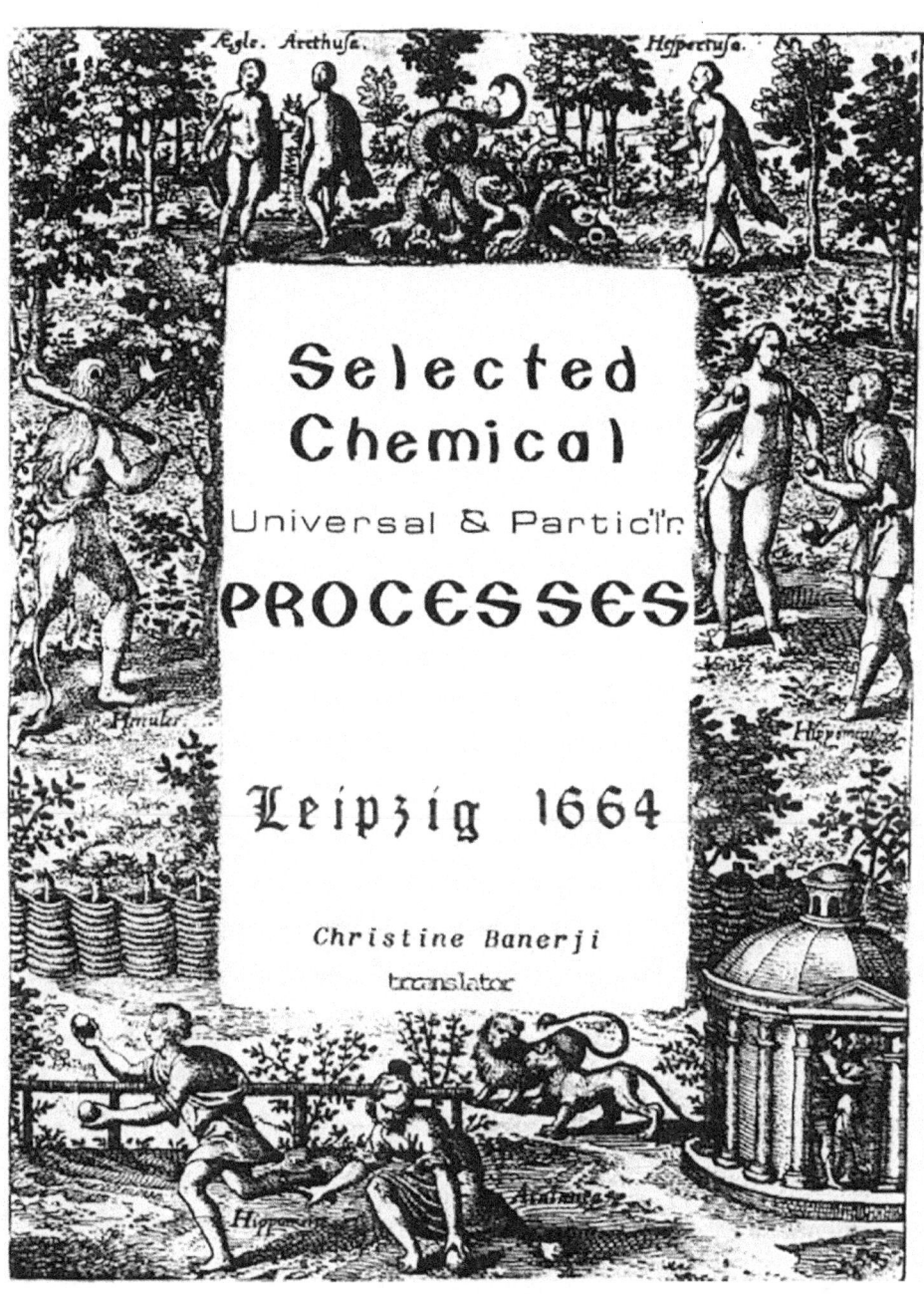

Selected
Chemical

Universal & Partic'r

PROCESSES

Leipzig 1664

Christine Banerji
translator

Foreward

Oh happy child, discoverer of this book and of this treasure of philosophical secrets! Praise be to you. For you will become a wise artist. For this book will tell you everything that is hidden between heaven and earth in the treasure-house of the wise art. Because it is fitting that all books on our art should have a forward or explanation, I have added this foreword here, so that you will know what is hidden in this book. You should know that great universal works of various kinds are to be found in this book. You will discover nature itself, as well as wonderful and glorious particular works, all kinds of graded waters, and fixations of ☿ with ♅, ♁ and ♂.[1] You will also find other kinds of particular works, of minerals, for example. You will find three little books which were found hidden in the walls of noble artists, as you will see. But make sure you keep the secrets, whoever you are, oh discoverer, for I swore a great oath to only communicate these secrets to one person after my death, and to forbid him to pass the secrets on, and to tell him to keep them for himself. But I found none who was secretive enough, so I asked GOD for

---

[1] You will find the book "Alchemical Symbols" to be helpful when interpreting the symbols in this book: Volume 21 in The R.A.M.S. Library of Alchemy. -pnw

advice, and it occurred to me that I should brick the books up, and then, he who was intended to find them would do so. So I hid them here in the year 1658 on the 9th. October, and got mason Casper of Kirchbach to brick them up with two pistols, and I warned him not to reveal the place to anyone. So I say to you, dear child, discoverer of this book, make sure you do not reveal the secrets to anyone. If you have no means to begin your work, choose a particular from this book and tell it to a knight in exchange for a piece of gold. Then work day and night until you achieve the work. When you have done so, pray hard and do not forget me and I will not forget you to GOD. Do everything to praise and honor God, and he will give you what you desire. Now that you have it in your hand you cannot fail, as long as you follow everything carefully. I have been zealous, as far as possible, in my efforts to describe everything correctly, so take care and don't wish me harm. For if anything fails, do not blame me. Blame only yourself if you should fail. If things don't work out as I describe, don't rush things, think of the proverb: Everything takes time. The farmer cannot harvest his seeds until the time is ripe. I won't spend much more time on my foreword, for you wouldn't like it. Turn to the book of wisdom and listen to what I have said.

I, Alexius, Freiherr von Ruesenstein, give this book to you, its discoverer. How and in what way I will give it to you, you will soon find out. Follow my descriptions exactly, as I will tell you. My dear son, you have everything here on the subject of the art of alchemy. In order that you might know my and your own heritage when you find it, I shall write it down from beginning to end. And if someone else discovers it, it will do him no harm either. You should know that my great grandfather - Peter Rueb - was a farmer on the Gapl, who became rich by selling medicines. He had one son, my grand-father, by the name of Friedrich Rueb. He was sent to the German School in Bischoflak. When he grew up and had come of age, his father married him to a burgher called Christina Weigerin. Soon afterwards his father died. He left my grandfather everything. He was a very wealthy man. They called him the rich Innkeeper of Kirchhoflack. Various people stayed at his inn. One day an alchemist arrived and stayed overnight at the inn. He told him about making gold. My grandfather must have been very interested, and they worked together to make a Gradating oil (which you will find in this book). However much ☽ is placed in this oil, as long as the oil covers it completely, will turn into fine ☉. My grandfather became so rich that he didn't know what to do with his money and his possessions. He had a son called Ferdinand

17

Rueb - my father - for whom he bought a house in Krauburg. He made him comfortable there and taught him the secrets of the art. My father kept his art a secret and became very rich. I was not brought up in the house, but was sent to board with a household in Labach from the age of twelve onwards. (I was the only son). I studied and completed the sixth grade at school before returning home to help my father in the inn for 23 years. So I was 41 years old when my father died. He gave me these writings on his deathbed, saying I should keep them a secret. For 12 years I remained unmarried and gave up the running of the inn. I was called the rich Herr von Rueb, and many people called me a goldmaker too. Soon after that I married a Frau von Hagenstein from Untercrain, Hagenstein. There were many rumours going round that I could make ☉. Then the leaders of the town questioned me: Everyone knew, they said, that my father was a rich man, but he had kept silent, and it had been assumed that he made his money from running the inn. But now, some said, Herr von Rueb was practicing the art of making gold. Would it not, they asked, be advisable to report this to His Majesty, the Most Gracious King? But I answered them: My dear Gentlemen, like every citizen and farmer I know that gold can be made by ploughing the earth. It is true that I am rich, for my father, my grandfather and I saved a great deal. Everyone knows I am my father's only son, and that I inherited everything from him. At this they were silent. I went to Untercrain to get some peace and built Hopenbach. I lived there for seven years. During that time I employed an alchemist who spied on me and went to the court to report me afterwards. I was then summoned to court, as you will hear.

I, the author of this book, am going to tell you the secrets of obtaining all the treasures of the world. I will tell you how I came to know them and where I got them from. You will learn all this. Once I received some ore from a farmer. I tested it as the ancients did. I found it to be true, I cannot describe the pleasure I got from doing this. Then I got several hundredweight of this ore from the farmer and worked on it for 4 years. I put aside the art which my father had left me because it was highly poisonous, and concentrated on the ore until I had got 20,000 florens from it. Then the farmer died and, because I did not know where to find the ore, I had to stop working on it. But now my enthusiasm for the art had been fired and I was driven by it to work day and right, reading and studying books written about the art by the ancients, so that finally, after much studying and practicing, I succeeded in making a particular tincture (without using what my father had left me), albeit a hard and coarse one. But through further experimentation I began to see more clearly, as you will read in the following report.

When I had learned the art I was greatly pleased, and couldn't imagine a greater pleasure. I went to the Royal Court in Prague where I had been accused by the alchemist I had sacked. There I reported to His Majesty Ferdinando III, and asked for an audience with him, which I received immediately, and I went to him. I told him I was experienced in this art and that I possessed a true Particular Tincture; that I had with me two pounds of the water I shall describe below, and I tested it in the presence of the King. When the King saw this he was full of admiration and said: You can do great things. You shall remain here in our protection, and he gave me a beautiful gilded chain on which hung a lovely coin with a portrait of the King. The chain weighed 1 1/2 pounds. I thanked him courteously and made to leave. But His Majesty said to me: Dear Rueb, I would like to say more to you. Come with me. He stood up immediately and led me into another room. Two of the chamberlains came with us, but he sent them away and said to me: My dear Rueb, now that we know you have knowledge of the art, I say to you: The Swedes are launching terrible attacks on us and we have very little gold. If you want to stay here and continue practicing your art, we will

protect you, and nothing will harm you. You will be
held in the greatest respect and authority. Then
I said: Yes, your Majesty, as you command. But I am
a commoner. I cannot mix with the nobility. Then the
King said: As you wish, I will make you a Prince. I
said: That would be too much, but I should like to
be a baron. The King said: Will you keep your name
or would you like to charge it? I said I should like
to charge it a little, my name is Herr von Rueb, but
I would like to be called Baron von Ruesenstein.
Then His Majesty called one of the chamberlains and
ordered him to prepare a patent of nobility in the
name of Baron von Ruesenstein. Bring it to me
immediately, said the King, so that I can sign it
and seal it with the Royal Seal. He also ordered
accommodation to be made ready for me in the court
where I could live with my servants. When everything
was completed I thanked him most courteously, and a
chamberlain took me to a floor not far from the
King's Antechamber, where I was to live with my
people. But before this, as I was leaving, the King
asked me to call him wherever I was practicing my
art, so that he could have pleasure in it too. I
always had the best food from the court, and 2 or 3
chamberlains always ate with me too. One evening I
asked a chamberlain, by order of the King, to go to
his Majesty and tell him that I was practicing my
art. The King came immediately to watch. That

evening I chopped up 100 double Thalers and turned them into the finest gold, the King was greatly pleased. I stayed there for 8 years and made over 100 tons of ⊙ for the King. Then the Swedes neared Prague and the King left for Vienna. There I begged leave of him and said I was getting old and could no longer leave it (my home) much longer. He did not want me to leave and I didn't want to teach anyone else my art, so I asked to go home for a holiday. Reluctantly the King allowed this. I made my journey back to Hophenbach. There I stayed for 4 weeks, then I bought an estate in Obercrayn in the name of Stermal. I built a house where I could live until my death.

Now, my dear son, discoverer of this treasure, you will find here many true things and the art of six learned adepts described in many different ways, just as they were told to me and as I learned them. But first let me teach you my art which helped me to build the church at Labach[2].

There now follows the beginning of his art, including the processes taught him by Schulz, one of the adepts. Then follows an explanation of the coat of arms in the church and a further story of how the adepts came to him and what they taught him. First, here is a Process found in an abbey.

---

[2] An area in the Südwestpfalz district in western Germany.

From the Manuscripts of Alexius, Baron of Ruesenstein.
Extract from the First Part.

His Own Process.

I took 1 pound of the mineral ♂ and 1 pound of the mineral ♀, and 1 pound of the mineral ☉is. I cut up these minerals finely and moistened them well with spirit of ☐ and left them in a cellar for 8 days. After this time the matter should have melted and dissolved. Then place it in a tall phial. The phial must be large enough for half of it to remain empty. Pour spirit of ☐ over it until only a part of the phial is empty, seal the phial well at the top with a blister and place it in horse dung for 8 weeks to keep it warm. After this time, take out the phial and pour the substance into a flask and extract the volatile spirit per balneum[3]. When nothing more rises from it, empty the flask into a well-coated retort and place a large receiver over it, sealed with luto. At first it should be heated very gently for 8 hours over a △ of the first degree of heat. Then apply the fourth degree for 12 hours. Then leave the retorts to cool down for 12 hours, so that the spirit forms in the receiver.

---

[3] "Per balneum" refers to a water bath, which works like a double boiler. -pnw

Then remove the receiver and pour the ∇ in the retort over beech ash or potash so that it rectifies and is separated. A receiver must, however be used, sealed with Luto, and must be driven with $\triangle$ and you will have a *Spiritum Rectificarum metallorum*. Pour this in a rather tall phial and when you have 8 Lots of it, take 1 Lot[4] of your chopped up gold or calx auri made with Regis, and when the gold is in the Spiritu, seal the phial well and place it in a sand bath (cupel) where the heat is so great that the ∇ boils constantly. The phial must be sealed with a bung: for every 4, 5 or 6 hours it should be removed to let out the steam so that the phial does not break. It should boil for 42 hours. Then let it cool down and place silver lamellas[5] or old coins in it. Leave them in there for 12 hours and they will have become spongy and bleached. Pour the remaining water off, take out the $\u263d$ you added and cupel it and you will see that your $\u263d$ has turned into $\odot$. You can continue to add more silver until all the water is gone. Another way, a far better way of making more menstruum follows:

---

[4] A Lot is a unit of weight measurement that was used in the Middle Ages. Recorded values for one Lot range from 10 to 50 grams. -pnw

[5] Lamella is Latin for metal-foil sheet. -pnw

Another Method of Doing the Same Thing

I took the three above-mentioned minerals and broke them up evenly into pieces the size of nuts. I placed them for 42 hours in a Johannis fire (which will be described at the end). Then I took out the substance and placed it in a pot or pan, poured water over it and left it to boil for a good hour. Then I filtered and coagulated it until it was not dry, but had become a thick liquor. I placed this liquor in a sugar glass in a cellar and placed a few wood shavings on top of it. In this way I made a fine ⊕. This vitriol I placed in a phial, as I reported before. Over it I poured a measure of one pound of Spirit of ⊡, placed it in horse manure for 8 weeks, as before, and followed the same procedures as before, but with the difference that this substance is purer because the ⊕ is boiled from the minerals.

After this he joined the 5 adepts on their travels.

Schulz's Process.

He took one pound of crude ☿ which he melted in a pan. When it had melted he added one pound of glowing ♂. He kept adding small amounts of ♋ until everything had melted. Then he left it on the heat for 1 1/4 hours, and poured it out into a Cone and let it cool. He took one part of this chaos and broke it up and melted it with one part of crude ♂, without adding any other liquid, for half an hour. Then he poured it out and let it cool down, tipped up the Cone and clinkers of the white King fell from it. You should keep these. I took the King, placed it in a pan, heated it until it flowed and added glowing ♂. Now and again I threw in pieces of ☉. I left it on the heat until the ♂ had been absorbed. Then I quickly poured it out and let it cool in the Cone. When it was cold I poured it out and I had the silver and white King without the star. Then I broke up and heated one part of this and two parts of crude ☿ for half an hour. Then I poured this out and let it cool down. When I poured the cooled substance out I found many clinkers and the King, as before with a beautiful star. Keep these two things. I took 3 pounds of the clinkers, broke them up and I added 1/2 pound of crude ☉, mixed everything together thoroughly and placed it in a rather large pan. I melted it, and when it flowed like water I poured it out and found a white King which is wonderful. Keep this and you possess the born Son who fixes all things. I took the King which was lovely and white, placed it in a pan and melted it. When it was very hot and melted, I added glowing ♂, adding pieces of ◐. I melted it until it had absorbed the ♂. When it would absorb no more I continued to heat it for 1 1/4 hours. Then I poured out a white King without a star. You should keep this. I took the clinkers, boiled them dry, filtered and precipitated them with distilled vinegar and a yellow ♄ fell to the ground. This I filtered, dried and stored. This ♄ is a subtle spirit which heats ☿.

I melted this King, but while it was melting I added as much of my ♄ as I could and threw coal dust in to make it melt faster. But because I had added all my ♄ I quickly poured the substance out and let it cool, after which I had a beautiful King. I roughly chopped the King. Then I poured 2 parts ☿ viv. into an iron mortar and poured boiling water over it. I stirred and ground it until a fine mixture was formed. I put this amalgam into a mill where it was ground day and night until all the impurities had disappeared. This took between 3 and 4 days. Then I dried my mixture, distilled the

☿ from it per Retortam and melted the ☉. When it had melted I poured it out and mixed it and found only a very little ♈︎. Indeed it was almost all ☿. Finally I distilled the ☿ off and placed it in a phial (the size of a small egg) and filled it with 6, at the most 7 Lots, sealed it well at the top, placed it in a sand cupel which I had filled with sand to a height of two fingers, to the same height as the substance, and with ashes up to the little tube. I heated it with △ at the second degree of heat for 14 days and nights. If you can see colours forming continue with this level of △. But if, after 14 days, nothing can be seen, increase the △ by one degree until it is completely black.

When the black phase is over a white colour can be seen. Then you can say: *Resurexit a mortuis*[6]. The △ should remain constant. The black colour will last for 7 weeks and then it will turn white, and white crystals will form in the phial. If you continue with the △[7] in the same way you will achieve the red colour in 8 or 9 weeks. Then I took one part of the red ♀ and ground it together with two parts of my well prepared ☿ until I had a pure amalgam. Then I repeated the procedures as above

---

[6] Literal meaning: raised from the dead. -pnw
[7] This symbol was missing in Hans's version. -pnw

until I had achieved the red colour again. This I did three times. When it had turned red for the third time I increased the △ by half a degree and the mixture became blue, and after it had turned blue it became green, and after green yellow, then brown, then grey and then red again. But when it was red I took 8 parts of this powder and one part of my ground gold and mixed the two thoroughly together. I half-filled a phial with them, placed this in a cupel and heated it for 24 hours at the 4th. degree of △. Then I removed it and transmuted the first thing which came into my hands. This is what you do: Take a piece of wax the size of a pea in your hand, press it out to the size of a Crown, take 1 or 2 grains of the tincture, place it on the wax, close it up again and throw on 10,000 parts of ☿ or ♄, whichever you prefer.

Ruesenstein's Process, Learned from his Father

Take one pound of ♂, one pound of crude ☿̇ (although minera[8] is better), two pounds of auripigmentum[9], 6 pounds of good Hungarian vitriol[10],

---

[8] Mineral. -pnw
[9] Orpiment, a deep orange-yellow colored arsenic sulfide mineral. -pnw
[10] A sulphate. This is often used to refer to sulphuric acid. -pnw

3 pounds of common ♀, two pounds of ① and grind
these substances together. Place the mixture in a
large glass or receptacle, pour on enough purified
spirit of ⊡ to cover it by a span. The glass must
be tightly sealed at the top and placed in a cellar.
Leave it to stand for 8 weeks, then remove
everything from the glass. Place it in a flask or a
different vessel and steam off the spirit of ⊡,
until it is dry. The △ should not be too strong, so
that the spirit doesn't smoke, but it should be very
dry. Then grind it to a fine ♂, pour it into a
fairly large retort, which should still be half
empty and △ it as follows: Take 3 pounds of the
mineral ♂ and a similar amount of the mineral ♀,
the same amount of ☉ and the mineral ☿. Chop these
minerals and grind them very finely. Then add 4
pounds of vitriol and as much ⊡ as is needed to
dissolve the vitriol. Pour the solution on the
minerals. Then make a sharp lye from equal parts of
beech ashes and four quarters of living lime. It
must be very acidic. Pour this over the minerals.
Put all this in a glazed pot, or another kind of
vessel which is suitable for ⊖s. Seal this well at
the top and leave it to stand in a cellar or a dimly
lit place for a long time. You can do this
immediately. After the allotted time has passed,

boil it in a tin kettle for a good hour. Then let it gently smoke so that it dries out properly. Then grind it and put it in an earthen or iron retort, place a large receiver over it and distil it for four hours over a gentle △. Then increase the △ as much as you can until the recipient is clear and nothing more drops from the retort. Then the retort should be cooled. Pour all the contents of the receiver onto the substances you ground before.

Finally, place the retort in an open △, placing a large receiver over it. Heat it for four hours in a gentle fire and then increase the heat of the △ as much as you can. (Note that the retort should be coated with good Luto so that it is not damaged.) Heat it fiercely for 24 hours without taking it from the fire and a sharp spirit will be formed which is completely red. Don't breathe in this smoke, for you will not be able to lute the retort in the recipient so well, and some will escape. Now the Gradated oil is ready. Then pour it into a glass phial, seal the top with a bung and place into Balneum in Digestionem. Let it digest gently for 14 days and nights, then take it out, add as much ☽ as you wish, let the △ cover only the cupel, and in 24 hours it will become pure ☉ without further fermentation.

This Belongs to Schulz's Process.

If I want to add ☿ as I told you, I do not proceed in the same way, but rather as follows: Before it is placed in the phial it should be amalgamated nine times with a ninth part of ☉ Vitr. Calcl., and distilled from it. But first, before the ☿ is removed the amalgam has to be ground until the ▽ poured over it flows cleanly. You should use warm rain water. This grinding will take about twenty four hours.

Here Schulz tells him that he has been keeping something from him.

Rusenstein's Oleum Vitrioli,

to be Included in the Above Process.

I put six pounds of vitriol in a large iron retort which I placed in an open △. When all the liquid had evaporated, I increased the △ gradually until no more smoke came from the retort. Then I let the △ cool, took the receiver off and stored my spirit in a glass. Then I also removed the ☉ from the retort, made it into a fine ♂, which was as red as blood. I poured this into a receiver, but the spirit which had come from it I placed in a glass retort and placed this retort in the sand. I also put a receiver containing the ♂, over it and gently distilled the spirit off. When this was finished I let the △ go out and took everything that was in the receiver and placed it in a glass retort. This I put into the open △. (The retort should be well coated). I placed the receiver over it and proceeded again as before. Then I removed the ☉ again, and proceeded again as before. I did this until I couldn't bear to do it any more: I repeated it seven times. One of my retorts was ruined, but its value was repaid a thousand times. The last time I ground the ☉ to a fine ♂ and placed it in a flask, poured all the Spiritum I had extracted over it (there remained only a quarter of the six pounds) and then sealed the flask well so that no vapour could escape. Then I put it in a warm place for 24 hours. Then I removed the flask and found a brown earth on the bottom. This removed and on the top I found my Oleum, as red as blood. I poured this through blotting paper into a flask and placed it over a condenser. But I placed the flask in a bath, allowed the phlegma to distill off, and my Oleum remained sweet and red as blood with no smell. Then I continued as follows: When I made a Spiritum, as already described, from arsenic and Auripigmentum, and from the minerals I have already described, I proceeded to the point where I said ☽ plates should be added, or old coins. Then I added this Oleum to about three parts of the above and one part of what I have just taught you. Then I let it stand for 24 hours in a warm place so that it could unite. Then I added the plates, and not only did they turn to gold within 12 hours, but when I removed them after 12 hours, they had welded together and I had gold which had too much colour. So I had to smelt a similar amount of ☽ to it before I could sell it as ☉. You should note that during this process, as soon as the receiver is full of a corrosive steam which stinks abominably, I let it escape through a little air hole. None of the corrosive substance remains, for it burns itself to earth.

Now Schulz will tell you of a Burgher's Process.

A Certain Burgher's Own Procedure

He took ☿⚖ made from vitriol and ⊖. He took one pound of this and mixed it with 8 Lots iron filings in a sugar glass, placed it in wet sand and the mixture melted and became watery. He distilled this liquid off per Retortam and put it aside. Then he took one Lot ☼ and poured 3 Lots of this liquid

over it and left it in the warmth until the ☉ was
at the bottom and had turned white. Then he poured
the clear extraction off in a little retort and
stored it. Then he took 6 Lots iron filings and
poured 12 Lots of the above water over them, and
extracted it in the same way. When everything had
been extracted he filtered it and added the
extraction to the Extracted ☉is in the little
retort. He distilled this to a liquor until the
menstruum was useless. Then he took 8 Lots ♀ filings
and amalgamated it in the following way with 16 Lots
☿. He heated an iron mortar, poured in a measure of
acidic vinegar, added the ☿ and ♀ filings, ground
them until he had a fine amalgam which he washed
with clear and fresh water to remove all traces of
the vinegar. Then he extracted this amalgam with the
same ☿ial water until the ▽ became coloured. He
poured the coloured ▽ over the Extracted ☉is and
the ♂ and distilled it nine times. The tenth time
he left the ▽ in it and placed ☽ plates in it. In
eight or nine hours they became ☉. You can do this
as often as you like, until the ▽ is gone. If you
distill the ☿ from the ♀ amalgam and cupel it, you
will get your ☽.

A True Particular

Take 2 pounds of ♀ crudum and 2 pounds of ☽. Grind them and mix them well. Detonate it as necessary in a mortar. When it has been detonated, grind it well. Add 4 or more pounds of potash. Place it in an earthen bowl in a brick oven, where it should be heated until it burns. When it is burned, dissolve it in common ▽ and boil it up to 9 or 10 times. Then take all the lye you can find and boil it up to two measures. Take 2 pounds of common ♁, grind it finely and boil it in the lye for 2 hours. Keep adding lye so that it doesn't boil dry, and so that there are always 2 measures of lye in the pan. After 2 hours filter it through paper, and boil it until it is dry. Dissolve it in rain water. (Use only as much water as is needed to dissolve the mixture) and boil the ♁ which was left in the filter-paper in the water for another hour. Filter the lye again and boil it until it is dry. Dissolve it again as before, and boil the ♁ in it again. After an hour filter it again. Do this 9 or 10 times. When you have done this it is good. Then you have a double Particular, a fine, liquid ♁ and a fixed, liquid ⊖. Then proceed as follows: Take one pound of ☿, dissolve it in 2 pounds of Aqua fortis.

Distill the Aqua fortis from it and you have a precipitate. Take one part of this precipitate and grind it together with one part of the ♀ which remained in the filter and place the mixture in a pan. Cement it for 24 hours under a gentle △ and then place your pan in a smelting △. Melt the substances together and you will find a yellowy metal, 24 Lots ☿ from one pound. Cupel this and you have fine ☽ and you will find in its marrow 4 Lots of ☉.

   Proceed as follows with the ⊖: Place one pound of ♀ filings in an iron mortar. Pour boiling strong wine vinegar over it and add 2 pounds of ☿vivum to it. Stir it well, and the ☿ will immediately attack the ♀. When this has happened place the ♀ and the ☿ in a glazed pot and add three fingers of dissolved, melted common pitch. Light the pitch from the top and stir well with an iron rod. Leave it like this for 8 hours, but keep adding pitch so that it keeps burning. After 8 hours you will have a sort of coagulant without having lost any of the weight. Then proceed as follows: Take 8 Lots of your coagulant, and 16 Lots of the ⊖ from before, melt the salt in a pan until it flows. Then make little balls out of the coagulant, the size of peas. Throw them in one after the other until you have used them all. Then increase the △ to smelting point. When you think the substances are melted at the bottom, pour everything out and refine it with ♄ and you will have made 6 or at the very least 5 Lots of fine silver from 8 Lots of coagulant. You will find 3 Lots ☉ in the marrow. This I found in truth.

Coats of Arms

There Now Follows an Explanation of the Coats of Arms, Which are to be Found in a Church in Labach, where the Art is Hidden and can be Learned.

In the first coat of arms you can see a green and a yellow lion like my own coat of arms. They are joined by a snake. Above them is a crown, and below them hangs the sign for ♀. In the middle there are two wings and two moons, and above these two stars. The background of the arms is black. The explanation is as follows: The red and green lion mean the minerals ♂ and ♀ and the crown means ☉is. The snake means that ▽ of ☿ comes from these things. That is why the sign for ☿ is found in the middle. The black background stands for the earth which separates these three. The wings say that the spirit of ☿ is very volatile and that you should store it carefully if you don't want it to escape. The moonshine means that ☽ turns to ☉ in this ▽, and that a star leads from them. This star, however, is the ☉ ex ☽ of the moonshine. The reason I searched for this in the three minerals is as follows: I had read much about nature, but what is nature? Nature is life. What is life? Life is the air. But what is the air? The air is a thing created by GOD to animate all things, and to keep them alive, and it contains all things. Without air nothing could be done. So I observed and found that the Corpus or Mineral is nothing other than a coagulated air which is made into chaos with the help of the earth. If it is putrefied it returns to its earlier nature, but is more fixed, trapped, exhausted and hungry and can be compared to that which is given to it. It consumes everything you give it, including ☉, but it gives it back 100 times over in the form of ☽. It can change the nature of everything. The coat of arms behind the Lauretanian Chapel means the following: A red and a blue lion can be seen, and a double snake and a double crown. The background is light in colour. This means: If the minerals are purified and only their souls taken out, they will make and give you double the amount, because previously the earth took a great deal for itself, and wouldn't let anything out. The advantage you have with this will be explained in my father's sweet ♋ of ♄, otherwise you won't achieve anything. I have also fixed ☿ with this oil.

A Process from an Abbey

A Process Also Found Among Herr Baron von Ruesenstein's Writings, Which is Said to Come from an Abbey.

Take 2 Lots of the true red Corpus of ☉ and 2 Lots of prepared white Corpus of ☽, and 12 Lots of ☿. Grind them together to an amalgam. Place this in a phial with a long neck. (The phial should have a rim in the middle so that you can place another glass over it and lute it at the bottom up to the edge when it is heated). Place the phial in boiled ashes in a dry place. Leave it to stand without △ for five days. On the fifth day heat it a little and replace the dung, the spiritual elemental △, and add 2 Lots purified ☿. Keep adding 2 Lots ☿ until the fifth day until you have added 9 further Lots ☿ to the amalgam. 9 parts of this make 36 Lots, which, together with the previous 3 parts of 12 Lots, make 48 Lots ☿, and 2 Lots red and 2 Lots white ☿ in the glass which you, with the help of

GOD will make into the corpus within 90 days. When
it has stood for 40 days it will turn black on top.
This blackness remains for 40 or 50 days, until the
Corpus is completely dissolved. Collect the black
substance from the mercury carefully and sensibly,
for this is the beginning of what you are looking
for. Place each one in a well-sealed glass (i.e. the
☿ in one and the black substance in another). Seal
the top of the glass containing the black substance
so that no smoke can escape. Place it in a
philosopher's furnace and heat it gently but
constantly for 8 days. The heat should be like
sunshine, otherwise the earth will become very
thirsty. Then it must be soaked in its own ☿ which
you have also stored, but watch the weight. Don't
give it any more than there is earth, preferably a
little less, for now dryness must have the upper
hand. Leave it to stand for 2 days and nights
without △, then place it in the △ again, so that
the ☿ can't escape. Leave it for 8 days, or until it
has dried out. Then dry it again as before and leave
it to stand for 2 days without fire. Then place it
in the warmth again. You must continue until all the
☿ (which has become 50 times as great as the black
substance at the beginning) is absorbed and
coagulated. Then the ☿ dissolves in this black
substance gradually and turns to ash and becomes one

and the same as the Corpus and is made pure and
white and subtle through washing and heating. While
the powder in the glass continues to multiply you
can keep adding more ☿, but always add less than
there is powder in the glass. You can also gradually
increase the △ because it can gradually bear more
heat. Don't hurry, take your time; just as you
needed 90 days to dissolve and moisten it, so will
you need 80 or 90 days to dissolve it in the
dryness. When the Corpus has drunk all its water and
is dry, heat it with a strong fire until it burns
and it will appear white. Be happy and give thanks
to God for if you want to ferment it with fine ☽ as
will be described below, one part of this will
change 10 parts ☿.

Now take one part of this white Corpus, 6 parts
of purified ☿, grind them well together and place
the powder in a phial. Place it, well-sealed, in
cold ashes and leave it to stand without △ and it
will turn black again. The substance is thick and
oily, that is why it is called Oleum Sulphuris. If
it has stood for 10 days without △ and you find it
has not turned black, the elemental △ in the
sulphur is too strong, the moisture is too weak and
it cannot overcome the dryness. You will have to
come to the aid of the moisture by adding more ☿

until the moisture overcomes the dryness. Your work is completed when a black substance appears on top. This will take 30 or 40 days without worldly △. But take care that you don't add too much moisture or the coagulation will take too long. Separate the black substance from it again, as before, and continue as before in every respect. However the △ should be increased now to speed up the coagulation. This will occur in 35 to 40 days. Then burn it again fiercely and it will turn white as snow.

Then take one part of this white ♀ and one part of purified ☽ (if, that is, you want a white tincture). Melt the ☽ until it flows and add the white ♀. Mix together.

Take 13 parts of this Corporis when it is burnt. It should weigh the same as the first powdered ☿, at least until the moisture overcomes the dryness and the dry Corpus dissolves without △. Now you have a Corpus, putrefied once more, soul and spirit joined, for they have been poured and melted together. Nature bears another Corpus, ashes and a black substance. Collect these again carefully. Burn them, soak them, wet them and coagulate them until their moisture, as you have learned, has completely consumed them. During this procedure, which will take you 40 days and nights to dissolve and coagulate, you will see wonderful colours. For all

43

the metallic colours will appear naturally. When all the moisture is dried up and coagulated, burn your Corpus well with worldly △ until it becomes white as snow.

Then take your wise and complete Corpus and divide it into two parts. Use one part for a white tincture and keep the other half for a red tincture. But because this tincture is far too strong and can transmute many thousands of parts, and because it would be difficult to transmute so much at one time, proceed as follows: Take ☿ which is 100 parts heavier than one part of the elixir you want to use, and wash this ☿ until it is clear and pure with salt and vinegar. When one hundred parts are washed place them in a pan over the △. When the pan is warm put one part of your elixir in it, stir thoroughly and everything will become an elixir. Then take another 100 parts of ☿ and one part of elixir as before, and everything will become elixir or powder. For a third time take 100 parts ☿ and one part tincture or powder and the ☿ will coagulate to a fine ☽. If however, through carelessness or other cause, your silver was not good enough, you will have to melt it again and throw a little tincture on it. This also applies to the red tincture.

How to Make the Red Elixir

Take one part of ☉ and melt it. Then take the other part of the elixir mentioned above, before it is fermented with ☿ and ☽. Heat it with strong △ and burn it for several days. When you have done this, add 9 parts of purified ♀, mix it well together and it will all become the same essential thing. But when you add these 9 parts ☿ you must do it without worldly △, and leave it to stand for a few days, and it will become a little moist and black. When this has happened, heat it with worldly fire more and more strongly and the black colour will disappear, and various colours will appear until in the end it turns into a beautiful white colour. Continue with the worldly heat and it will turn red. Burn it well and you will have, with GOD'S help, the true and complete elixir, which will transmute ☿ and all metals to true ☉. Do this as in the same way as when you were using the white elixir; always use 100 parts ☿ to one part elixir. I tell you, if you want to transmute common ☉, this elixir can transmute 1,209,984 parts to good quality ☉. This is why your ☉ and ☽ are better than natural gold and silver.

If you want to dissolve your elixir to increase its quality, do the following: Take the last of the powder or tincture which was increased twice with one hundred parts of ♀. Take some of the red or the white, whichever you wish, and pour lots of ☿ over it, so that the moisture gains the upper hand (all this is done without △) and so that the top turns black. Collect the black substance as you have been taught, and give it its ▽ to drink. Using worldly △, coagulate it, burn it well and give it its ferment, that ☉ or ☽, as you have been told. Burn it further and add 9 parts of ☿ to it. Then let it stand without △, mix it and it will unify. Burn this very well as you did before, and as you have been taught, and you will have made an elixir again. Increase this again with 100 parts of ♀, and repeat this three times with 100 parts, as you did before. Once again you will have ☉ or ☽. You can stop repeating this process in 50 days. It can go on for all eternity; all you need to do is repeat it as you have been taught.

Virtues of the Elixir

On the Elixir's Great Virtues in Illness and the Maintenance of Good Health

These 2 elixirs, the red and the white, before they are increased with the ☿, have the power to heal all illnesses. If an illness has lasted a year, give the patient an amount the weight of a mustard seed and he will recover in 12 days. If however the illness has already lasted for many years it will be cured in a month if the patient takes an amount the weight of a mustard seed every seven days. Hermes says; If you take a mustard seed's weight of our elixir on seven consecutive days your grey hairs will fall out and black ones will grow in their place. If this powder is held to the rose of a woman in childbirth, she will soon enjoy the birth of her child without complications.

Tinctura ex Sulphur Solis, or Sun-Dust

Take one pound of dust from a room in which the sun shines and in which white cloths are spread. Pour two pounds of rain spirit on it and place it in the △. Let it stand for 24 hours and the spirit will absorb the ♀. Then filter the Spirit, distill it, and take 1/2 Lots of it, 4 Lots rainwater ⊖ (for there is only very little of it) and 1/2 Lot ☉. Grind it well for 6 hours in an iron mortar and pour over the rainwater spirit to enable the ♄ and the ⊖ to dissolve. Then grind it for a little while again, filter it, distill off the Spirit, save for a little ☊, and if you proceed like this you will soon be able to transmute.

Universale Gualde ex Stercore & Urina Aalium

Addito Iron & Gold & Mercury

I extracted a ⊖ from earth dampened with urina aalium, very often found under a well. I poured spirit of wine over it so that the ⊖ could dissolve, extracted it again, and I have made a nice ⊖. Then I took 8 Lots Reg., ♂, and poured it seven times through fresh ♁. I melted 1 Lot ☺ with it, milled it finely and took 18 Lots of my ⊖, ground it with all my might in an iron mortar until a black foam had formed on top. I poured a little S.V.R. over it to dissolve it, and ground it for a while longer. Then I distilled it and a yellow ☉ remained. I took one pound of this and added it to one pound of ☿vive and ground it until it became a

48

black ♂⃰. Then I distilled the ☿ again per Retort, ground it to an amalgam, washed it clear with S.V., placed 6 Lots of it in a phial and proceeded to boil it off, as you already know. Then I fermented it with a ninth part of common ☉, and in this way I changed 300,000 parts. I poured S.V. over the ☺, from which the ☿ had been extracted, I let it dissolve, filtered and coagulated it, and thus I could use my ⊖ again. Who, dear brother, would believe that such a noble thing could be in animal or vegetable matter? I wouldn't have believed it myself, but I discovered in nature that it was the case. For, if the farmer doesn't fertilize his fields well, he will not bring forth many fruits. But what is the reason for this? It is for no other reason than that the fruits which he plants need a ⊖ , and no common ⊖ either, but a purified ⊖ which has been putrefied twice, once inside the human being or other animal, and once in the air or similar place. The longer the dung lies and decays, the better the farmer likes it. For, he can do more with this dung than with dung which is not decayed. The reason for this is that it becomes richer in ⊖ through putrefaction. So, my dear brothers, you can see that the true Menstruum is nothing other than a putrefied ⊖ . After this Schulz said to Gualdo: Dear brother, I have learned this: I have found it in Theophrastus's writings, in the May dew and in snow, in rain ▽ and similar things, these things are all the same in nature, for after putrefaction they contain a wonderful ⊖. Gualdo said: It is true that the true menstruum can be found in them, but it is too simple to think that nothing more is needed than these things. It is false because nature is in nothing other than in ☉, and this is only the menstruum which lets the ☉ out. Then Fauermann said: But there is no ⊖ in our Regulus, but rather only ♃ and we transmute it. Then Gualdo said: I would tell him who doubts that there is a ⊖ in our Regulus the reason why I need the Regulus, and I would like to solve this puzzle for you at the same time. If someone wants to make a tincture, he should now that he must destroy the ☉ and purify the ☿ and amalgamate it. I have done all the correct things to do this. I added one lot ☉ to 8 lots ♄. I melted them together, and when I poured them out the ☉ was open until I tried to extract it again using a cupel. Then I took these 9 Lots ♄, added 27 Lots of ☿, poured ▽ over it, and ground the amalgam until it gave off a black substance. With this black substance the Terra ☉is, ♄ and ☿ go, and only the ⊖ and ♃ remain in the ☿. This mixture contains

three things, as everyone writes: ☿, ♀ and ☿. The
☿ consists in ☿, the ♀ in ♂ and ☉. For the ♂ has
the characteristics of ☉, and contains a ☉ like ♀
which clings to the ☉ and remains there. If one of
you should object, and say that the ♀ is in ♂ and
the ☿ in the ♂, for the ♀ of ♂ is the same in the
amalgam ☉is and the ♂ carries more ☿ than ♀ in
itself. My answer to this would be; the ♂ might not
have a ☉ like ♀ in it, which should not be
discarded, but if you want to use it, you will have
to know how to separate it from its ☿. And when it
is separated you will have to know how to open it up
to introduce it into a metal. You know very well
that ♂ contains little ♀ and ☿, but that it has a
pure ☿. But this ☿ can be used more particularly
and universally in our art than ♀, for ♀ is brittle
and coarse, and ☿ is agreeable, sweet and liquid
and gentle. This is why the ♄ won't have it, because
it carries with it coarse ♀. If someone were to say
to me: what kind of ♀ does the ♂ have which is the
same as ☿? It is a pure and poisonous thing which
is harmful to metals as well as people, for it makes
all metals brittle and wild. I would answer that
person: Why is the ♂ poisonous? Because it is open,

and all metals are dangerous when they are open and volatile. But if they are fixed they become harmless as before. You can test this by doing the following: Take 8 Lots ♂, add 1 Lot ☉. Melt them together and pour the substance into a Cone, and a King will fall to the floor. Place this on a Cupel, melt it, blow it, and the ☉ will have a better colour than before. For what reason does the volatile ☿ cling to the fixed ☉, why does the ☉ colour become better? The reason is none other than that the ☉, is fixed, and if something which is volatile is added to it, that which is volatile fixes with the fixed ☉. This happens because if you add volatile matter to fixed matter, the volatile matter becomes fixed or the fixed matter volatile. Schulz said: All ⊖ which are not volatile are good. Gualdus answered Schulz: Yes, all ⊖ are good, but first they must go through the process of putrefaction. Only a few ⊖ which are already compact can be made to putrefy, otherwise they can easily decay because they are still in their prima materia. Rain, dew and snow are the best, they are all equally good, but those made from ⊡ and Stercus vary. They have nothing from the sun and the air, as rain and dew have. But I think it is possible to make a stone from every single thing in the world. Using vegetable matter, I proceed as follows: I let the vegetable putrefy until worms appear. Then I wash it out and collect the worms which have the Menstruum ☉ is in them. I place the worms in a well-sealed glass container in the sun and wait until they have turned to water. I distill this ▽, I make the ⊖ and Spiritum, and proceed further in the same way as with other substances. Everything is one, whether it comes from heaven or earth, all you need for our art is a putrefied ⊖. Now I would like to tell you how Wilhelmus and Albertus Magnus took their Menstruum from the sea. They did the following:

Menstruum ad Tincturam ex Mari

They took ▽ from the sea and put it in a well-sealed glass, which they placed in the sun in the summer, or in a warm place or in horse dung in the winter, where they left it in this natural heat for the whole winter, or until the ▽ at the bottom had many feces and looked a whitish colour. This indicates putrefaction. Then they distilled off the ▽, poured it over again, and repeated this 9 times. Then they distilled off the Spiritum. They put the ▽ to one side. They ground the ⊖ with an eighth part of ☉ until they could see that the a.a. had been extracted and that the ☉ looked destroyed. Then they poured the rectified Spiritum over it and ground it for a while so that the ☉ could dissolve with the ⊖. Then they proceeded as normal.

Schulz said: This is what I heard and what I have taught Ruesenstein: They took sea ▽ and coagulated it and dissolved it. Then they fermented it with an eighth or ninth part of ☉ and transmuted 30,000 parts. Gualdus said: Dear brother, you have misunderstood. After it was putrefied he coagulated it and then he distilled the Spiritum using a condenser. Then he dissolved it again, but not with a different ▽ , but using its own Spiritum. He didn't throw the Spiritum away, but used it to open up the ☉. For it is very strong and penetrating. He made the ⊖ spiritual through putrefaction, and through repeated distillation he managed to get a lot of ⊖ volatile. This ⊖ volatile opens up the ☉ ex fundamento. So you see, dear brother, one person uses ♀, another uses sea ⊖, a third ⊖ commune, a fourth uses rain ▽, a fifth uses dew, and so on. But if you look at them all clearly, there is no difference between them. I have even used S.V. in this way: I left it to putrefy in horse dung or in the sun, I tied it with a blister and put it in a pot and covered it with another so that the blister was not damaged. Then I distilled off my Spiritum universum, and proceeded normally. Koller said to Gualdo: I know an alchemist who took 4 pounds of mineral ☿ and ground it to a fine powder. He poured rainwater over it, (collected from a thunder-storm) he filled a glass two-thirds full with it, sealed it and left it in horse dung for a good half year. Then he distilled off the Spiritum until moisture came off. Of the ☉ he took 8 Lots and poured 2 Lots Spiritum over it. He ground it until it was clear and placed it in a warm place for 24 hours. Then he filtered it, distilled the Spiritum off and he had a yellow ♀, which was the a.a. ☿is. This is the same as the a.a. ☉is. He poured the Spiritum over 8 Lots materia, and extracted it again until he had a new materia. When there was no more ♀ in this materia, he calcined it over a fire and stirred it until it was ashen grey. Then he poured the afore-mentioned Spiritum over it, the same weight as before, and extracted the ⊖ from it. He took one part of this to two parts △[11], poured the same amount of Spiritum over it (which had been rectified and from which the Phlegma had been extracted), dissolved it to an ⚭, placed it in a phial, so that 3 parts of the phial were still empty, seal it hermetically and

---

[11] This is the symbol used in the R.A.M.S. text. It is an error. -pnw

55

proceed, using common △ as my brother told us. When it was in the third colour, he fermented it with a ninth part of ☉ and transmuted 20,000 parts of all metals. Gualdus said: I can believe this because the ♀ in the ♂ is more red than that in ☉, if you can only find it, and there are very few people who can do it properly. I tell you in truth, I would rather open up ☉ and extract its aam[12], than ♂.

Gualdus said: Another time I made a stone from ⊡ in this way: He drank only wine and let his own urine putrefy in a sealed vessel. Then he distilled the Spiritum through a tall flask. With this he extracted the ⊖ and extracted the Spiritum from it so often until the ⊖ was white. Then he took 1 Lot of Miller's ☉, 4 Lots of this ⊖, ground them together for 4 days and nights until it became as black as coal. Then he took 12 Lots Spiritus, washed the mortar out with it and placed everything in a flask. He sealed it and placed it in a warm place for 24 hours. Then he filtered the Spiritum, distilled it, and he found a yellow ⊖. The black Terra ☉is was left in the filter is useless. He poured the Spiritum over it and distilled it, and did this 20 times. When the Spiritus ⊡ became weak he took some more fresh Spiritus ⊡. Then he took

---

[12] "aam" is exactly what appeared in the R.A.M.S. text. -pnw

his yellow ⊖ and added 1 Lot of this to 100 Lots ☽ or ☿ and it was changed to pure ☉.

Particulare cum Augmento Perpetuo ex Vitriolo

Take green Hungarian ♑, dissolve it in ▽ and coagulate it until it forms a skin. Crystallize it and keep doing this until the ♑ is as blue as ultramarine. Then distill the Spiritum off, boil out the ☺ so that all the ⊖ is taken out, take 16 Lots rain ⊖ or May dew ⊖ or March snow ⊖ or putrefied ⊡ ⊖, and 8 Lots of the raw powder, grind them in a mortar for 24 hours, place this powder in a glass and add 1 pound of the same Spiritus as the ⊖ is made of and leave it to stand in the warmth for 24 hours. Filter the Spiritum until it is red. Extract it per alembic. Then take 4 Lots fine ☽, laminate it thinly. Take 1 Lot of the ⊖ I taught you about before, make S.S.S. in a pan which has been chalked, lute the pan, place it in a circle of ▽ and cement it for three hours. Then melt it together and you will have fine ☉. You will find this, you can rely on it, said Gualdus. But if you want to take it further, take the same powder again, the same ⊖ and Spiritum, and when the Spiritus is red, pour it over the ☉ coloured powder again. Do this as often as you wish, and each time it will transmute higher, but only ☽ - no other metal.

Particularis Tinctura ex Arsenic.

Take 1 pound of Florum, (sal ammoniac) 16 Lots alb. arsenic, 16 Lots ☿ and grind them. Place the mixture in the cellar until it is wet and fatty. Extract the ▽ and heat it with sublimating △, so that the materia can sublimate. Then break up your materia or ☺, place it in the cellar again, leave it to stand again until it is wet, and place it in a flask. Pour the ▽ from before over it and proceed again as before. Repeat this until the arsenic & the ☿ turn to stone in the glass. Then take 8 Lots of this and 2 Lots of fine Miller's ☉. Grind it in a glass mortar until the ☽ can no longer be seen. Pour it into a small coated flask, half of which should remain empty, seal it well at the top with a bung of lime, and place it in a sand bath near the ground. Cover the flask with three fingers of sand. At first the △ should be gentle, but keep increasing it until it glows and you hear crackling in the flask. Then let it cool down and you will find a white stone which you should chop up. Add 1 Lot of the stone to 1 pound melted ♀ and it will turn to silver. You can also transmute ☿ with it.

Another One of Ruesenstein's.

I took 1 pound of arsecic alb., and 1 pound of common Arcanum duplicatum and ground them well together. I made an ♋ of ♃ from 3 pounds of ⊖ of ♃. I made a dough from these things and left it to stand in the cellar for 14 days. Then I distilled it per retort until all the moisture had gone, and increased the △ until no more steam came off. Then I took the ☉ and chopped it up finely. I poured ▽ over it again and left it in the cellar for 4 days. I repeated this 4 times. The fourth time I made a dough with the ▽, and covered ♀ lamellas with it, dried them gently, heated them until they glowed, washed them in vinegar and cupelled them to make fine ☽. I can also do as follows: I grind the ☉ and cement it with ☽ S.S.S., for 3 hours and melt it. This makes just as fine ☽ too.

Schulz Tells Another.

Someone took ♄, and calcinated it to ashes. He took two pounds of this and ground it with 1 pound of nice white arsenic. He poured over ♋ of ♀ until it was completely soaked through. He put the sugar glass containing the substance in the sun until it was dry. Then he proceeded as before. He repeated this 12 times. Then he placed it in a pan and cemented it for 4 hours in a circle of ▽ . He added it to 2 parts ♄ in a cupel and he had made fine ☽.

Another One.

He took one pound of ♄ and melted it and heated it in a pan. Then he added as much white prepared arsenic as the ♄ would absorb, and poured it out. He had made a vitrum, he ground it, put it in a sugar glass and poured the following ♋ over it: He took 1 pound of a.a., ♀ ⊖ and detonated it with the same amount of beech ash, the same amount of ⊕~ ⊖. He mixed these ⊖s with three parts of potash, burned it well, boiled it up with rain ▽, filtered and coagulated it, burned it again with 3 parts potash. He repeated this three times. Then he left it to turn to ♋ in the air. With this ♋ he wet the ♂ to make a dough. This he dried in the sun until he had ♋, the ♋ must be made of 3 pounds of ⊖ or 1 pound ♄ and arsenic and then he placed the mixture in the cellar until it was damp and then dried it in the sun. He did this three times and then cemented it for 4 hours before placing it in a smelting △. He added several pieces of ⊖ and the metal was precipitated. Then he poured it out, cupelled it and he ended up with as much ☽ as there was ♄ at the start.

Another, Told by Gualdus.

He took 1 pound of ☉ & ♀ a.a., detonated it and made it into an ♋. Then he made an Aqua fort. from ℞ and allum. He took 1 pound of arsenic alb., and poured the Aqua fort. over it and left it in a warm place for 24 hours. He extracted the ⊖ from the ☼, took 1 pound of this and mixed the two ⊖s together. He burned them, dissolved and coagulated them, made them into ♋, mixed the two ♋s together, i.e., this last one and that which was made of ♀ and ☉, and poured it drop by drop into the Aqua fort. with the arsenic. He put it, well-sealed and constantly simmering, in a warm place for 24 hours. Then he poured it over 5 pounds of ♀ plate, which he left to dry in the sun. He cemented it for 6 hours, luted and melted it, and had his ☽.

Particulare cum Arsenic ad Copper in Silver.

A certain adept ground 1 pound of arsenic alb., and placed it in a sugar glass. Then he poured the following ♋ over it: He took 1 pound of common ♃ and mixed and ground it well with 2 pounds Tartar crude. He placed it in a pot with 3 pounds of well ground potash. He burned it well and then he boiled it out with Aqua fort. until no acidity was left in it. He filtered and coagulated the ▽ and took 2 parts of this salt and 1 part ♃ and repeated the above process 3 times. The third time, when it had boiled dry, he hung it in the cellar in a little sack with a glass bowl underneath and the ⊖ flowed out as an ♋. He took as much of this ♋ as he got from 3 pounds of ⊖ and added it to 1 pound of arsenic and proceeded as follows: When the arsenic had been ground he added ♋ to it in a sugar glass until a dough formed when it was stirred, and then he put the glass containing the substance in the sun until it dried. Then he ground it to a subtle ♂. He placed this ♂ in a glass and added the same amount of ♋ as before. He repeated this until there

was no more ♂♂ left. Then he ground the mixture to a
♂, put it back in the glass, placed the glass in a
damp cellar and left it until it had become very
wet. Then he poured it into a flask, sealed the
flask at the top, and placed it in warm horse dung
for 4 weeks. Then he took it out and placed the
flask in a sand bath, covered the flask half way
with sand and heated it, first gently and then with
a fierce △ until the flask glowed. He left it in
the △ for 4 hours and then let it cool down. In the
flask he found a white stone. He ground this stone
to make a subtle ♂. He took 4 Lots of this ♂, 1
pound ♀ plate place it in a pan and cement it for 4
hours and he had ☽ instead of ♀. Gualdus said: I
will tell you why this happens, but first I will
tell you how to prepare May dew. Take as much May
dew as you can get and pour it into a kettle. Boil
it for a good while and pour it into a little barrel
and cover it with little planks of wood. Leave it to
stand under the roof until it begins to stink and
decay. Then proceed as I told you, but take care
that the ▽ from which you wish to extract the ⊖ or
menstruum has been boiled or simmered before it is
left to putrefy. Without boiling it you will get a
Spiritum, and it will also putrefy without boiling,
but it will be weak and bad. However, if it has been

boiled first, it decays more easily and gives a greater volume of Spiritum.

Gualdus said: Dear brothers, do you know why these processes use arsenic? It comes from nature, for ⚇s are natural things which bring about everything which is raw, for ⊖ is the most fixed thing which can be found. This is why, after everything has been burned, decayed and destroyed, a ⊖ can still be found in it. This ⊖ is the first and oldest thing or earth. But there are many ⊖s, is one good as the other? No; some have vegetable natures, others mineral natures; but once they have been burned, dissolved and coagulated, so that they lose the ♀ in their essences, they are all the same. You should understand this about the fixed ⊖s. But you should also know what volatile ⊖s are, which we need ad lapidem. These ⊖s exist and float about everywhere. They are nothing, invisible and incomprehensible, not sweet, not sour, not bitter. Where can they be found, if you can't see or touch or perceive them? They can be easily found everywhere, in every ▽ there is a ⊖, but it is not easily perceived. But if it is boiled and putrefied, then you will find a ⊖. And this is the volatile, invisible ⊖ which is only made visible through our art. All philosophers write about it and say our materia is invisible and imperceivable; it is nothing, not vegetable or mineral or metal. There is no ⊖, but through our art it is made into a ⊖. And this ⊖ contains everything I have said.

After this Ruesenstein left the six men, who signed with their own blood that they would strengthen these writings. Ruesenstein travelled to Sternohl from Salzburg. He asks the discoverer of these writings to keep them secret and to use the works not in his but in GOD'S honor and for the good of those who come after.

Extract from Two Little Books

which the 5 Adepts gave to Baron von Ruesenstein.

Here is the ▽ and the △, believe me dear friend, I will show and tell you everything, so that you will know it too.

Chaos is made as follows: Take 1 pound of ♂ and iron, as usual. Make a Regulus from the Blackmahl (?) ♀ and ⊖, make the Regulus ♀ic with Blackmahl and common ⊖. Boil out the clinkers, precipitate the ♀ with distilled vinegar and put it in a Regulum with the ⊖. This is the field, the ☿ is the seed. Then make a Regulus again and mix it with ☿ and grind it until it gives off a black colour. Then

65

distill the ☿ off per retort, mix it with the ☉ again, and the ☿ is ready. Now, take 2 parts of this ☿, 1 part ▽, suphurei, mix them and place 1 Lot of this mixture in a phial. Place it in a △, one degree of heat, and in six weeks you will have a black colour which will last for 8 weeks. Then you will find the philosopher's ⊖ or crystals. Break open the phial and collect the crystals from the red earth. Put the red earth in the phial again, place it in a △ at the third degree of heat and in 14 days you will find even more crystals. Take all the crystals, add them to the mixture from the first regulus and with ♀ precipitated in a phial, no more than 5 or 6 Lots and put it in the above-mentioned place, and it will become black again. This won't take long. If the first phase was completed in 13 or 14 weeks, this will be completed in 5 or 6 weeks and you will find crystals which you can multiply as often as you wish. They will always transmute. If you imbibe it or incinerate it, it will increase, but it will not transmute as much as if you multiply it. You can continue until 1 part transmutes 30,000 parts. Prepare the ☿ to imbibe or incinerate it as follows: Take 6 Lots ☽ and 1 pound of ☿ and mix them, grind them and dry them. Place the mixture in a phial and heat it first gently, then fiercely. In

4 weeks you will have a red preparation. Remove it and place it in a stone mortar. Grind 4 Lots volatile ⊖ of ⊡, 1 ½ Lots volatile ⊖ of ♄, and take white, sublimated ♓ and grind it. Then the ☿ will become alive, and nothing will be left but the ☽ and a little black ♂. When you have half a pound of crystals, take 2 pounds of ♀. Divide it into 7 parts and imbibe it. If the white crystals are imbibed after multiplication you will find a red ♂. If you incinerate this ♂, again it will turn blue. If you want to use it *ad medicinam primae ordinis*, you won't need to imbibe it again. Instead, take 1 Grain in a little glass of wine, leave it to stand for 12 hours, give it to the patient to drink and he will be cured. Pour 1 part ♀ onto the blue King, grind it in a stone mortar and place it in the △ at the second degree of heat until it turns green. This is how you imbibe it. After the green colour it will turn brown. After imbibing a yellow King appears. The last imbibing leads to a grey King. And from this comes a red King which is 7 measures high. These are the seven imbibings. To ferment, take 1 Lot of our fresh ♃ic regulus and 1 1/2 Lots ☉ chalk (calx auri), melt them together, grind the mixture, and take 9 Lots of this ♂. Grind it in a glass mortar and place it in a glass phial in a △ at the

third degree of heat for 4 days and nights. It will become a yellow ♂ which is the highest tincture. The amalgam must be ground, before it is used for as long as it takes for a blackness to be given off. This occurs in 14 days and nights. Make sure that 3 parts of the phial remain empty. The oven should not be too hot, the little stems which have grown should not fall over until you have achieved the ☽ Philosophorum.

A True Fixatio Mercurii and Transmutatio

Copper in Lunam

Take 1 pound of ☿. Dissolve it in 2 pounds of Aqua Fort. Add 1 measure of S.V.R., and place the well-sealed glass in a warm place for 24 hours. Black feces will be seen. Filter the ▽ and the feces will remain in the filter. Extract the moisture and you will be left with the ♁ of ☿. Pour S.V.R. over it and distill it off again. Do this 8 times and then filter the ♁ through blotting paper. Leave it to stand for 8 weeks, sealed, in warm horse dung. Then take copper plate, heat it, and wet it with this ♁, place it in the fire and it will turn to ☽. You should fix the ☿ in the same way. Put 1 pound of ☿ into a coated flask, place it in the fire so that the ☿ begins to rise. Add one grain of the ♒︎, leave it to stand for 24 hours in the fiercest fire, cupel it and you will have made a ☽ fixam.

Gradatio Lunae Fixae.

Take four pounds of iron filings and extract the redness with 2 measures of vinegar. Place it in dissolved ⦶ and yellow feces will fall to the bottom. Leave it to stand for 4 days and nights in a warm place, filter the essence and the feces will be left in the filter. Distill it until a skin forms at the top, let it cool down. Filter it until it weighs 24 Lots, pour 8 Lots ♒︎ of ♁ over it per campana and it will bubble up. (It should not be all poured over at once). Filter it again, place it in horse dung for 14 weeks, filter it for the third time and place the clear liquid in a glass. Gently dry it and you will be left with a red, sweet ⊖ which has the consistency of wax. Add 9 Lots of this to 1 Lot ☉ chalk (calx auri) grind the two together and place the mixture in a phial in the sand. Heat it fiercely and the ⊖ will become liquid and the ☉ will be dissolved. Leave it flowing for 4 days and nights and you will find a bleached stone. Add one part of this to 100 parts ☽ which should be too hot to touch. When the ♂ has been added place the ☽ plate in a glowing fire until it too glows. Then not only the ☿ial Luna will be graded, but the common Luna too.

In the ☉ you will find a ♄ which is white. Dissolve it in distilled rain ▽ until it is white. You can work with it dry or wet, as Basilius taught: It is a pure spirit of the air which contains ⊖, ♀ and ☿ which cannot be separated. Take 4 Lots of this ⊖, place it in a phial, hermetically fixed, bury the balls in ashes which are hot, but in which you could hold your hand for 2 or 3 minutes. In a short time you will be able to see a red ⚇ and the substance at the bottom will be black. This ⚇ will coagulate again and in 3 months it will turn white. The ⊖ will turn red within 7 or 8 weeks and must then be fermented. Take one part of the red ⊖, one part Miller's gold and place them in a phial again as before. They must be augmented and fermented three times. But the third time don't augment it, just ferment it, but just as before, one part ☉ to three parts ♀. It must be placed in sand and river ▽ must be added, and your tincture is ready. Transmute it in the following way: Heat 2 pounds of

♄ or ☿, place 2 grains at the most of the tincture in wax, throw it onto metal, and it will form a yellow skin and become the finest ☉. This is the wet method of doing it, as Basilius describes: I take 16 Lots of the ⊖, place it in a phial, pour over one part filtered rain ▽, so that the phial is half full and add 1 1/2 Lots fine ☉. I lute another phial over it and heat it with △ until it boils, until it is blood red. It shouldn't boil too strongly though. Then I distill off the phlegma, ferment it and augment it as I did in the dry method. After augmentation I pour the phlegma over it again, place the well luted phial over it again and boil it again in the red substance. I do this 3 times as above, and ferment it as before. This is how Basilius did it, and it is but one substance, not 2 or 3 as many say and write, but they have misunderstood.

There Now Follows a Description of

Ruesenstein's Portrait.

My name is Alexius von Ruesenstein,
In my hand I hold two snakes,
Which you will recognize,
They are called ☿.

Aries is also here,

He is called Martis,

For he is also my image.

I am portrayed as a prophet,

Between the four elements,

Which are shown as angels.

The phial is my sceptre,

I use it when performing

The secret work

Which Schulz taught me.

My procedure is not the same as his.

But he was the one who taught me.

I distill my ▽ with arsenic,

But Schulz dissolves his in ☿.

He did teach me this too.

But I do not do it this way,

He tried to show me that his way is better than
mine,

And I want to show him that mine is better.

So don't think I'm lying

I have buried no lies,

Remember me with an Our Father and an Ave Maria.

May God have mercy on me,

Now and forever,

Blessed is the most holy trinity.

The Process spoken of here is that which is reproduced above and which is called Schulz's Process.

Conversations with Five Adepts

Now Ruesenstein Tells How the Five Adepts Came to Him and What Sort of Conversations They Had Together.

Late one evening 5 men arrived with the excuse that they had got lost on the way to Labach, and that, because night had fallen, they could not go on. They asked me for shelter for the night. When I saw their servant with the message I was unnerved: I thought they must have come from the Kaiser's court and that they would recognize me. But when I came down to receive them I realized that I knew none of them. So I welcomed them and asked them to share my humble abode. I led them to my room and asked them for news of the world. While we were talking one of them kept looking at the corner of the room where there was a door to another room. Then he took down some $\triangledown$ and looked at it and took me aside. He asked me what it was. I answered him; it is $\triangledown\mathrm{o}^{\!\top}$. He said; it is good. Now Baron, you must know who we are. We have come here because we have often heard speech of you. People say you are a great and experienced alchemist. All five of us are experienced artists and adepts, and we have come in order to learn something new from you, and in return we will show you our art. When I heard this, my heart felt so light I thought I was not on earth but in heaven. I asked him their names. The man said; my name is Schulz, the tall one is Fauermann, the third is Koller, the fourth Fornegg and the fifth Monte Schider. Baron, you will learn wonderful things from us. I thanked him with all my heart and Schulz and I rejoined the others. Fauermann said to Schulz; what is happening? Is everything alright? Schulz answered; yes, perfectly, and all of them congratulated me. Schulz transmuted a whole pound of mercury into gold before my very eyes. I had the evening meal prepared. We became close friends and I promised to travel with them for a while. While we were eating Schulz told us the story of a Swiss farmer who had jaundice whose doctor advised him; take one pound of ♂ filings and place them in a glass. Pour about a quart of strong Swiss wine over them and put the glass in a warm place. Stir it 2 or 3 times a day and when the wine has turned black, drink 2 small glasses of the substance every day, morning and night. The farmer followed his advice and recovered. The farmer was clever; he thought he would be able to help others in the same way. So he prepared some more of the substance but used S.V., instead of wine. The S.V. turned red, he strained it through cloth and poured it into a glass. Then he poured more fresh S. V. over the ☉ and it turned red again. He repeated this process again and again and made a large amount of tincture. He distilled the S.V. off, then poured it over again, and it turned red again. He repeated this many times. It so happened that another farmer in his village became ill and he gave him a spoonful of his tincture and the man recovered within two days. An alchemist visited the farmer and told him to distill the S. V. from the tincture. Then he placed 2 ducats in a pan and heated them until they glowed. He poured the tincture over them and a yellow powder was formed. This he added to the lead and it turned into gold. Then he told the farmer; now you know something you didn't know before. Use it to do good for the poor.

Fauermann told us a historical poem: An eagle joined with a dove. The dove had young ones who turned into golden apples. The meaning? The eagle is my Mercurius, the doves are my Regulus, which, through nature, give birth to the philosopher's stone.

Schulz: I make a chaos from ☿ and ♂ and let the mixture cool down. Then I have in my possession the true matter, there is no question about it. I pour it out and melt it with ♂, I repeat this until

75

I receive a sign, A beautiful white virgin appears. She is the ☿'s food. He must die with her, but he does not decay. He becomes pure and can join with all metals as a tincture can. Brothers, I teach you this.

Fauermann: Now I would like to tell you that my art consists purely in dragon's blood. For I use the clinkers which others throw away, in my art. I melt them together with ♄ and ☽ and pour them through a Casting Cone. I find a red man inside who joins the ☿. But how do I tempt him out? I boil the clinkers dry. I filter them and precipitate them and then I cannot go wrong. I'll tell you more another time, we can talk again about what happens next.

Koller: I go on a long journey but only climb a small step. ☉ is a hard substance, difficult to unlock. Before I manage to draw out its soul I have to undergo much hardship, I have to find the ☿ and the one must overcome the other. For ☉ and ☿ must become one, so that they can bear a son. The son can penetrate all things in order to change them into something beautiful.

Fornegg: Now let me tell you what I do, what I have discovered which will bind all metals. But it is difficult to make. I don't dry the ☿, nor do I keep it trapped in a vessel, for it is always searching for cool air. Its greatest pleasure is to fly away, so great care must be taken to keep this bird still. I'll tell you more about these things on our journey.

Monte Schider: I know something, although I don't often talk about it, but I will show you that I can make the Lapis. I can only make a small amount. But I would like to pass this knowledge on to you. I know the mineral ♄ very well. I call it the mineral of ☉. I bind it with ♂, I find in it the bond of the ethereal man that will join with the ♀, I amalgamate it with ☿ and then triturate it, and then I have in my hands the beginning of the great work. I will reach the end too when I have done all that is necessary. We will speak of this again if God is willing.

As we were walking around my pond Schultz said: we are walking around a pond which is very wet. If however, the philosophic fishes are given very little to drink, they will start to decay and stink, and no one will enjoy them, but will tread them underfoot.

Koller said: This is my art: I mix ☉ meum or ☉ cicinat. or 2 Lots of ☿ and 8 Lots of ♀. I grind them and mix them until the gold no longer sinks to the bottom and until almost the whole mixture looks like ☿. Then I boil it for two hours in sharp vinegar. This boiling removes the acidity and the black colouring from the ☿. Then I grind it again. The longer it is ground, the better it becomes. When I can see that the water coming from the amalgam through the grinding is no longer black I stop grinding and dry my amalgam well. Then I pour not more than 6 Lots into a phial. The phial must be well coated with good lime, and must be big enough to leave exactly 3 parts empty when the mixture is poured in. It must be 2 measures away from the cupel at the very most; the bulb of the phial must be completely covered with sand. You can also use filings, in fact they are better. To begin with, a gentle △ should be applied for 10 or 12 hours so that the moisture can slowly steam off. Then heat of the 3rd. degree should be applied until the cupel glows brown at the bottom and the ☿ circulates round and round without stopping, and you see, it turns black as coal, then white, then red as blood. The fire must become stronger and stronger. You can do this in any oven. When you have seen the highest red, remove the phial when it is cool and you will find a lovely red ♂. Take one part of this and 2 parts ☿ ☋, and mix the two together. Repeat the above procedure with the △ again. This will putrefy rather more quickly than the other. The first takes about 8 weeks, but this only 6 weeks. Continue as before until the substance is red. (It is no good if it turns white immediately or if only the red colour is seen.) When you see the red colour, let the phial cool down and pour out the red powder. You will find corporeal ☉ at the bottom. Grind this finely and mix it with 3 parts of ☿viv. and do with this mixture exactly as I taught you before. Then add 4 Lots to ♂, i.e., if you have 2 Lots ♂ then add 4 Lots of the mixture to it. Do exactly as before, place it in an oven and make it turn red. Then increase the △ and it will become highly red. Then remove the phial from the oven and you will find a vermillion coloured ♂. Take 8 or 9 parts of this powder and grind it together with 1 part calx auri and place this mixture in a smelting crucible. Put it into the glowing part of the fire and as soon as you see a fiery flame coming out of the pan, remove it from the fire and you will find a sulphur coloured powder. One part of this can tinge 1000 parts. But if you want to augment the quantity and the quality of this substance, then do with the mixture as I told you before. I prepare the ☿ as follows: I mix 9 parts with 1 part calx auri and then distill it off again. I do this 9 times and then it is ready. In this work the most important thing is that the ☉ goes with it to putrefaction, it must become ♀ with the ☿.

Schulz: It is a strange thing that no tincture can be made without ☿. Wet or dry, the ☿ must be present. It is said of Basilius Valentinius that he used ♁ instead, but this isn't true; his ♁ is ♀ ♎ made with ♁. His writing is clear, but few understand him. He takes the ♁ and melts it in a pan. He presses 1 pound of ☿ through leather for every 8 pounds of ♁, and stirs the mixture until the ♁ is dry. Then he takes 2 pounds of ⊖, mixes everything together and ends up with a green sublimate. He takes 12 Lots of this sublimate and mixes it with 6 Lots of iron filings. He places the mixture in the cellar until everything turns to ▽. Then he draws it off per retort and finds a mercurial ▽. He pours this into a flask and removes the phlegm from the ▽ in a bath. He is left with an ∞ which he calls ⌒☿. He takes all of this and adds it to a ninth part of calx auri. He places it in a bath until the ☉ has dissolved completely. Then he places it in a phial and leaves it for 8 weeks in a △ and the ☉ joins with the spirit and becomes a penetrating spirit. He pours 1 pound of ☿ over this ∞ while it is still warm and a single drop of ♃. The ☿ is one of the wonders of the world, nothing can be done without ☿. Every adept in heaven and on earth uses ☿.

I said: I have also used ☿. I have fixed it, but with great difficulty, and I have turned it into ☉ and ☽. This is how I did it: I mixed 1 pound of ☿ and 1 pound of ♀ and heated it in the △. I poured this mixture through 2 parts of crude ♁ and I had made a lovely regulus. I added more ♀ to this until it turned silvery white and had a star on it. I ground together 1 part of this ▽ with 12 parts ♁ and placed it in a triturating mill and ground it until a blackness came off, and pure ☿ appeared and the ♁ kept the ♀ with it. Then I added an eighth part of fermented ☉ or ☽ and ground it again until I saw that the mixture was thick and hard and as white as snow. Then I placed 12, 14 or 16 Lots in a phial and fixed it in 22 weeks. I placed the glasses directly on the cupel and heated it fiercely so that the substance was always circulating. When it would climb no more, I removed the phial and found a red precipitate. I took one part of this, and 2 parts of ☿ animati. The ☿ is animated just as I taught you before, except that it is boiled from its mass per retort. I make the following mixture from this: I

take an iron mortar, heat it and pour in the ☿ with
the precipitate. I grind this until an amalgam is
made. I heat this amalgam as before and it turns
black and then it becomes a metal. This metal is
fixed. 24 Lots from 1 pound remains on the cupel.
The fourth part is ☉. But if I take the metal, file
it, amalgamate it again with 3 parts ☿ do everything
as before, then I find at the end that it is more
fixed and that the third part of it is ☉.

　　Schulz: I made it like this: I took 4 Lots
☽, mixed it with one pound ☿. I ground this mixture
until I saw that the ☽ was unlocked. I placed the
mixture in a rather large phial and let the ☿
circulate until it would climb no more. Then I let
my △ go out and removed my phial and I found a
lovely red precipitate. I put this in a glass phial
and poured sharp distilled wine vinegar over it. I
left the phial to stand in a warm place until I saw
that the red precipitate had become a brown earth.
Then I filtered my vinegar and boiled it off in a
condenser. At the bottom I found a pure ⊖. I kept
this ⊖, I took the brown earth and mixed it with 3
parts potash. I boiled it off in an iron retort and
I got lots of ☿ which was very fresh. I kept this.
Then I took the calx with which I had mixed the
precipitate, distilled it, melted it and I purified

my ☽ and became rich in ☉ at the same. Then I took 1 part ⊖ and 2 parts ♀ and ground them together. I placed 6 Lots of this amalgam in a phial and applied it in the first universal work I told you about. Within five weeks it turned black, 5 weeks later it was crystalline white and in a further five weeks it was red. I extracted this red ♂ with vinegar again, as before, but I didn't get any ♀ from it, I got ⊖ instead. So I made another ☿: I took 1 pound of 🜍 and melted it in a pot. I pressed 1 pound of ♀ through leather into the sulphur and stirred it for a quarter of an hour with a spatula. I let it cool down. Then I broke open the pot and found a brown lump. I ground this finely and mixed it with 3 parts potash and 1 part iron filings or iron scale, and distilled off a lovely ♀ per Retort. This ☿ is so greedy, it runs towards the ☉. Then I took 1 part of my ⊖ and 2 parts of this ☿ and mixed them together and heated again as before. This time it putrefied even faster, and turned white and red more quickly too. After it had turned red it became blue immediately and within 3 weeks it had gone through all the colours. Then I fermented it as I did the first one and tinged 12,000 parts. I prefer this one to the first. If you want to begin something ad

Universale, then use this one. I know it works because I have tried it.

Then we talked of how the ancients would make their Lapis without △, without glass and without ☿, for they took sea ⊖, and dissolved and coagulated it in the sun. Then they used the warm sand at the edge of the sea shore which removed all the impurities from the ⊖, and brought it to its first perfection. All they did was ferment, then they tinged. I have read about this and have found out that the ⊖ becomes a liquid after hundreds of dissolvings and coagulations and that it melts in the sun. So it must be just as capable of penetrating and it is true that it dissolves the ☉ and enters the metals as long as it is dissolved many times. But it cannot have its ingress as quickly. Then Schulz said: I believe that you can even make the Lapis from vegetable ⊖ because the ⊖ is not the tincture, but just the menstruum which opens up the ☉ and enters it. The ☉ alone can transmute and the complete secret of the art consists in preparing a menstruum which opens up the ☉ ex Fundamento. This is what must be done. Some break their heads in vain - I did at the beginning - I thought there was only one single substance which would show me the way to nature, but there are many thousands. Now, when I am working, I simply take the first plant I come across, and extract its ⊖ and make a menstruum. For it is well-known that ⊖ comes from the air in everything, and the air is nature itself. For when I extract the ⊖ from the plant, I am not taking the nature of the plant, but the nature of the air which was given to the plant, and which nourishes and keeps the plant alive.

Schulz said: If you want to burn ▽ you must do it in the following way: cut up the plant finely and let it rot in a well-sealed glass placed in horse dung in a warm place for three weeks. Then distill it and it will give the ▽ its nature. Burn the ☉, extract the ⊖, and put it in the ▽ and it will retain its strength. The extract must decay first too, then it is squeezed and soaked.

Fauermann once said to me on our journey: I once knew someone whose tincture was made of pure ☿ and tinged 20,000 parts. He took Hungarian Mineram ☿ and broke it into nut size pieces, put them in the secret △ of which you already know. He left them there for 4 days and nights. Then he took them out, steeped them in lye, filtered and coagulated the substance with pure rain water and he found a white ⊖. But because this had been dissolved with ▽ the water was like blood. He dissolved the ⊖ one hundred times, coagulated and filtered it repeatedly until a red ♋ came up through the condenser like blood. He poured 8 Lots of this ♋ into a phial and added 1 Lot of calx auri, put it in a Lamp oven at the same heat for 4 weeks until the ♋ no longer rose but remained fixed with the ☉. Then he removed the phial and found that the ☉ was completely decomposed and that the ♋ appeared to be a little yellow. Then he took 1 pound ☿, warmed it in a pan, and let a single drop of the ♋ fall onto it and immediately all the ☿ turned into the finest ☉. I saw this with my own eyes.

Schulz said: I believe you. When people come to me and ask me to teach them something, I tell them; choose a menstruum which dissolves ☉ and you will have a lovely tincture. Then he told us the Burgher's Process.

Ruesenstein said: I too have found an ancient process using spirit of ☿; make a chaos or regulus from 16 Lots of crude ♄, 8 Lots of ♂ and 8 Lots of ♀. Grind it finely and pour over as much menstruum as it can extract. When everything has been extracted, pour the extraction into a little flask and distill it to a liquor. Then take 2 ducats of gold and do the same to it. Then take 2 Lots of auripigment and extract it too. All these solutions must be poured into one flask. Place a condenser over it. Boil off the ▽ then pour it over again. Do this nine times and the ninth time boil it dry and you will find a yellow powder. Add 1 Lot of this powder to 4 Lots of ☽ in flux, this is how it is refined.

Koller: I half-filled a large phial with 6 pounds ♁~ comune and added 3 quarters of well putrefied May Dew. I sealed the phial with a blister, buried it deep in horse dung and left it for 12 weeks. Then I opened the phial. The inside was black and stank terribly. I boiled off the moisture until the substance was like dough. The water which steamed off was black and lots of volatile gas came off in the beginning. I kept this gas and I will tell you what I did with it later. I put the substance from the phial in a retort and distilled it using a large receiver, first gently and then more fiercely, until no more spirit came from it. Then I put the spirit in a sealed glass, placed the receiver over it and boiled it off very fiercely. In this way I got half a measure of blood red ⁑ in 24 hours. I poured this ⁑ over 16 Lots of ♂. It quickly ate up the ♂. When I saw that everything was dissolved, I put the substance into a retort and distilled it first gently, then fiercely, with a receiver. I kept the white ⁑ I got from this. I ground the ☉ to a ♂ and it was as red as blood. I put it in a phial and poured over half a measure of S.V.R. I sealed the phial tightly and left it in a warm place until the S.V. looked like blood. Then I filtered it through a covered funnel and drew off the S.V. in Balneo so that it became oily, so that the ♀ or tincture was not dry. I poured the distilled S.V. over the ☉ again. I did this again and again until a redness came out. Finally I boiled the extraction dry and kept it. I took the white ♋, and added 3 Lots to 1 Lot of ☉ and let it dissolve. Then I took 4 Lots of ☿, dissolved it in 8 or 9 Lots of this ♋. I poured these two solutions over my red extraction and distilled it per Retort. (The distilled ♋ can be used again and again.) I took the ☉ out of the retort and ground it finely. It was the colour of blood. I cemented 1 Lot of this with 8 Lots of ☽, I melted them together and I had the finest ☉ instead of ☽. Many people extract the tincture from pure ℞ and achieve the finest tincture.

Schulz didn't agree with Monte Schider because he said he was too rough and underhanded. Once he was talking about this and he said: I will teach my brother to make a fine tincture: Grind 1 pound of the mineral ♄ to a fine powder and melt it in a crucible. Then take 1 pound of old rusty nails, heat them until they glow and add the melted mineral to them. Apply a smelting △ until the nails have all been consumed by the ♄. Then pour it into a mould. Take 1 pound of this substance and 4 pounds of ☿ and make them into an amalgam. Melt the ♄, then pour in the ☿ and stir it with a little stick. An amalgam will be formed which you should put in an iron mortar and pour ▽ over it. Grind it for 2 hours so that the ☿ dissolves the ♄. Then dry the amalgam, put it in a wooden container through which an iron wire has been pulled to enable the amalgam to be sieved. Boil it until it has all turned to ♂ and there is a lot of liquid ☿ in there. Take the living ☿ out and you will find a grey powder. Put this in a phial. You can put one or two pounds in the phial - it won't do any harm. Heat it with the 2nd. degree of △ and the substance will go through all the colours within 8 to 12 weeks. When it comes to red, remove it and heat a mortar. Just before the mortar starts to glow, pour the red powder in. For every pound add a pound of the ☿ you ground before, and grind them together until the ☿ is no longer visible. Then place it in the glasses again and do the same as before and it will go through all the colours. Do this 7 times. The seventh time, when it has turned red, take 8 or 9 Lots of it and grind it with 1 Lot of Miller's ☉. Place the mixture in a crucible, place the crucible in a glowing fire and leave it there for 2 hours. The crucible must not start to glow. You must stir it all the time with an iron rod. When ☽ or another metal is poured into the crucible, pour 1 Lot of your tincture over it for every 300 Lots. Although it is good, it tinges poorly. The reason for this is that the materia is not strong enough to unlock the Corpus ☉ ex Fundamento. He also used ☿ sublimate: Dissolve 1 pound of ☿ in aqua fortis and precipitate it over the heat with ⊖. Sublimate it. Take 12 Lots of this sublimate, add 2 Lots Miller's gold and place them in a clear flask. Sublimate the sublimate again, take the ☉, grind it together with the sublimate until it is fixed in Fundo and has become a red Vitrum. Then add 1 Lot to 20 Lots of ☽ in flux and it will be refined. I prefer this last one because the sublimate attacks the ☉, with force because it is a ⊖ and it must unlock it.

Fauermann said he took a regulus instead of the mineral ♄. He did the following: He took 1 pound of ☿ and 1 pound of ♀ and made a chaos from the two. He added flux and 1 pound of both ♀ and Ⓞ, 16 Lots of ⊖ and 16 Lots of white ♁. He ground these things finely together and added 8 Lots of this mixture to the ♅, and when it had melted he poured it out, broke up the clinkers and let the King flow out. He put 8 Lots of the powder in flux again for as long as it would stay in flux, and until the clinkers were no longer yellow but came off white from the King. This King is very volatile. He took 8 Lots of it and 24 Lots of ♀. He poured the King into a crucible which must not start to glow because the King can melt even in a gentle heat. When it had melted he gradually added the ☿ and stirred it with an iron rod. He made a fine amalgam. I ground this, he said, in an iron mortar with ▽. When it is dry again I put it in a container and sift it until it is a powder and until most of the ☿ is live. I grind this powder in a mortar and grind the ♀ off. I fill a phial with 8 or 9 Lots of this powder, put it in a sand cupel in a Fauler Heinz (a furnace) 2 knife blades from the cupel (the bowl must be completely filled with sand) and I heat it with a △ of the second degree of heat. The materia goes through all the colours. When it is red, I imbibe it with an equal part of the ☿ from before. Then I heat it again and it goes through all the colours again. I do this as often as I want to; at the very least it must be done 3 times. I can distill it until I have enough. When it turns red for the last time I ferment it like this: I take all the red powder I have and grind an 8th part of calx auri or Miller's ☉ with it. I cement it for 5 hours. I am left with a yellow powder which will tinge 1000 parts. If I repeat the imbibing process more often, however, I can even make 1 part tinge 3 times 1000 parts. The only difference is that ♀ is used instead of ♄. ♀ is sometimes called ☉. I knew a man who could make 38 pounds of ☉ from every hundredweight of ♀.

Fauermann said: There was a man who took 1 pound of ♑︎, and 1 pound of ⊖. He smoked them and detonated them as is usually done with ① and other things. Then he dissolved the mixture in ▽, filtered and coagulated it and placed it in a well-sealed flask in a warm place with 1 measure of S.V.R., and the S.V. extracted the redness. He filtered this red spiritum vini and distilled it in a bath until it was oily, then he took the ☉ made from equal parts of ♑︎ and ① and added it drop by drop to the oily substance until it had absorbed it all. Then he placed it in a cool place and crystals formed. The ▽ should be poured off and the crystals left to dry. Then dissolve it in rain ▽ and coagulate it again. When you have dissolved and coagulated it 12 times, take 20 Lots of this ⊖ and 2 Lots of Miller's ☉ or ☽, if you prefer, grind it together with the ⊖ and place it in a bowl in the cellar so that it melts alone. Boil off the moisture and melt it again in the bowl. Repeat this 4 times and the fourth time, when the ▽ has been boiled off, dissolve it with the ▽ you have boiled off and coagulate it. Repeat this 20 times, the more often the better, and increase the heat of the △ and the ⊖ will turn into red glass in the flask. Remove it (it is perfectly transparent), grind it finely and grind 2 Lots of ☉ as I told you before with 20 Lots. Put it into a cement container or a crucible, cement it for 3 hours in a rather fierce △, but don't let the crucible start to glow. Then take it out and tinge, as I have described before.

There was a man who took 1 Lot of ♄ ♆, 2 Lots of arsenic and ground the two together, placed them in a driving wheel and placed it in a glowing fire. He let the arsenic steam off until it was completely smoked. But the △ was only gentle otherwise the ♄ would have started to melt and could not have been used any more. When it had stopped smoking he took another 2 Lots of arsenic and ground the ♄ with it. He did this 12 times. When this had been done, the ♄ was liquid, like wax. He put this in a phial, placed it in a Lamp oven with one light lit, and it turned black, white and red. Then he fermented it again with an eighth part ☉ in a phial and tinged 1000 parts with it. Another man made the Lapis from pure ♂. He ground it with water in a stone mortar until it had turned to powder. He dried it in the sun and then triturated it for 14 days and nights. Then he cupelled off the ☿, put 8 Lots of the powder in a phial and it changed through all the colours. Then he imbibed it three times with a sixth part of ☉ and was able to tinge 2000 parts.

Shultz said: A certain gentleman discovered this tincture: He poured a half measure of vinegar over 16 Lots of ☽ cornua, and left it to digest for 14 days and nights. Then he filtered it. He poured another 1/2 measure of vinegar over the substance left in the filter and repeated the process until he could extract nothing more. He knew he had reached this point when he poured a little ⊖ in the vinegar and it did not turn white. When he could extract nothing more, he boiled off the vinegar (ad ma) added ⊖ and the ☽ sank to the bottom. He filtered it with cold ▽, dried it and poured S.V. over it. He left it for 8 weeks in a steam bath. Half a measure of S.V. is used for every 16 Lots of ☽. The S.V. is made as follows: Take 1 measure of S.V.

94

which has been rectified 5 times. Add 8 Lots of Flores ☽ comune, and distill it in a flask. Put what is sublimated from the Sal armoniack back into the retort and distill it again. Repeat this 5 times and it is ready. If the putrefaction occurs in Balneo, filter the S.V. left in the filter. You don't need this. Distill the S.V. until it is dry and you will have a tincture of 8 parts of ♀ or ☿. If you wish to augment your tincture, add freshly prepared S.V. to the putrefied S.V. and add the same amount of ☽ cornua again. You can do this as often as you wish. If you repeat it twice it will tinge 16 parts, if you do it 5 times it will transmute 24 ♃.

Koller. I made ☽ cornua and proceeded in the following way using ♄: I put each in a separate glass and poured a quarter of sharp vinegar over every pound. I left them to stand in a warm place until the vinegar no longer boiled up from the bottom of the glass. Then I filtered the vinegar and distilled it until it was dry. Then I poured it over the ☉ again and extracted it. I repeated this until all except a very small amount had been dissolved & extracted. When the vinegar had been extracted for the last time, I washed it with ▽ cold. Then I took 2 Lots of the ☽ and 6 Lots of ♄. I ground them finely and poured them into a flask with 18 Lots of strong aqua fortis. I left the flask in a warm place while they dissolved and then I added 18 Lots of S. V., sealed the flask with a blister and left it to stand for 24 hours in a warm place. This is how the corrosive essence was precipitated; I filtered the water and distilled it and I was left with a red ♋︎ which I poured into a phial, sealed it hermically, placed it in a B. for 8 weeks and the ♋︎ decayed. If ♀ plate is painted with this ♋︎, left to dry and then melted, you will have your ☽. Fix the ☿ like this: Heat it and then drip ½ Lot ♋︎ on it for every pound of ☿, cover it and place it in a smelting △, and you will have your ☽. It can also be boiled dry per Retort and mixed one part to 2 Lots ☿ or ☽ then you will have fine ☽.

Fauermann. This is my other work: I make a regulus as I have often told you, melt it until it flows, and then add 18 Lots of flux. I leave it in the heat for half an hour and then pour it out. I knock off the clinkers and melt it again with the same amount of flux for the same length of time again until the clinkers turn yellow. The flux is 1 pound each of finely ground ♄, ☉ and ⊖, and 16 Lots of ♃. I take 8 Lots of this regulus and 24 Lots of ☿ and I grind them in an iron mortar and rub them with ▽ cald, until a blackness comes off them. Then I boil off the ☿, melt the ☉ and make an amalgam again of this and the ☿. I repeat this 7 times. The seventh time I put 6 Lots of this crude amalgam in a phial in the sand, heat it gently until the moisture has all evaporated, and then heat it with the second degree of △ until it turns red. Then I imbibe it with a ☿ Rectified which is made in the following way: I take a precipitate of ☿, revive it with aqua fortis with 3 parts potash. I take 2 parts of this, grind it in a warm mortar until it becomes a soft amalgam. Then I melt it again. I do this 3 times. Then I ferment it in the phial with an 8th. part ☉ and it tinges 10,000 parts.

Fornegg said: I knew someone in Turin who taught me to make the following particular: He melted common ⊖ and dissolved it with rainwater, coagulated and dissolved it up to 100 times. Then he took 8 Lots of ☽ cornua and between 5 and 6 Lots of this ⊖. He ground them finely and mixed them and placed them in a sealed flask which he put in a cupel and heated it with the 3rd. degree of △. It increased in volume. He powdered this and coagulated it and refined it 12 times. He added 1 Lot of this to 1 pound of warm ☿. It became fine ☽ and the 5th. part ☉. He said that it was advantageous if the rainwater was decayed and foul and then distilled. Then the salt dissolves properly, because it has a ⊖ of the air in it.

Fornegg continued: There was a man who made a wonderful medicine using a regulus of ♂. He discarded the first clinkers. But the second - which had flowed to the regulus from the Nitro as long as it was coloured - he kept, rectified and precipitated. From this he got a yellow ♀. Then he took foul and distilled rainwater, poured it over the ♀ and it extracted the redness. Even when he distilled it, the ♀ remained red. He took 2 Lots of this and 1 quarter of Miller's ☉. He cemented it for 4 hours gently in a crucible so that the ♀ didn't burn and he had a red glass. He broke this up and ground it to a powder. He uses just one grain of this to cure fevers, rabies etc.

Ruesenstein: I made it in the following way: I took 1 pound each of ♂ and ♂ and made a chaos from them. I poured the chaos through 2 parts ♂ crude, and melted the King which fell out with ☽. Mine was like brother Fornegg's except I mixed the clinkers (the first ones, not the second) with 1 pound ♀ and 16 Lots ☽ I melted them and added 1 part of my red ☽. When the two had melted together I kept adding coal dust. When the coal dust no longer burns the ⏀ you have to take it quickly out of the △ otherwise the ⏀ will absorb the sulphur immediately. Then I break off the clinkers and find that the King has become 2 or 4 Lots heavier. I repeat this until I have no more red ⏀. Then I take 8 Lots of the regulus and 16 Lots of ☿ ex Cinnabar nau., I amalgamate them, grind them finely and distill off the ☿. I melt the regulus in the retort, break open the retort and then melt the regulus with a little ⊖. I amalgamate it again with the ☿ from before. I repeat the whole process 7 times. The seventh time I fill a phial with 6 Lots of the amalgam, place it in the sand and heat it at the second degree of △. I let it go through all the colours. When it is red I imbibe it with 2 parts of the ☿ and I made it this way. Then I continue as I did three times ago. Then I ferment it with an 8ᵗʰ part ☉ and tinge 600 parts.

Theophrastus's work: He took ☿〜 with aqua fortis mixed with 3 pounds ♃ and one pound ⏀ for every pound. He sublimated it to a green sublimate. He sublimated this 3 times with (an equal quantity of) ⊖ and poured spirit of ⊡ over one part of this sublimate and put it all in a phial. He sealed the phial and buried it for 6 to 7 weeks in horse dung.

After that time the ♀⚊ is all extracted except for a very small amount. This he filtered and removed the phlegma. An ♋ remained. He sublimated this per retort with a large receiver and his ♋ became clearer and some feces remained behind. He took 8 Lots of this ♋ and added 1 Lot of calx auri to it. In this way the ☉ was dissolved. He half-filled a phial with this solution and put it in a Lamp oven. Within 8 or 9 weeks it turned black and white and red too. Then he added the same amount of ♋ again. He repeated this three times and fermented it with the sixth part of ☉.

The dry method of doing it is as follows: He dissolved ☿ in ♋ of ♅ which is available from the smelter's shop. He rectified it three times. Then he washed the precipitate and when the ▽ had been distilled off it was white. But it soon turned yellow like wax. Then he refined ☿ 7 times per retort. He took 12 Lots of this and 6 Lots of ☿⚊ and ground it in a warm mortar until no more ☿ was to be seen and he had a good amalgam. Then he washed it with fresh water and filled a phial with 6 Lots of it. The phial must remain three parts empty. The phial he placed in a fixing oven in a sand cupel, 2 knife blades high in the sand. That is to say; the bowl is well covered with the sand. It should be heated with the 2nd. degree of △ and within 6 weeks it will have turned black, within a further 6 weeks it will turn white, and in six weeks again it will turn red. Then he imbibed it with 2 parts of the purified ♀. He did this three times and fermented it with the 8th. part ☉ in a pan. He tinged 30,000 parts. When it is in its first red form it is called Medicinam Primae ordinis and it can cure all of the main or universal conditions.

Schulz also taught us Gualdi's method. He makes a regulus with one pound each of ♂ and ♂ poured through 2 parts ♂. He pours this through 2 parts ☿ 7 times. Then the regulus is yellow. Then he grinds 1 Lot of ♄ with 8 Lots of this regulus and amalgamates it with 3 parts ☿ as I do. He distills it and melts the ☉ and then amalgamates it again with the ♀ from before. He does this 7 times and puts 6 Lots in a phial just as I do. After this he does exactly as I do. He also has a S.V. which he makes like this: He pours S.V. into a glass and puts a bit of cotton wool into the glass too. He lights the cotton wool and catches the smoke in a cooling condenser. In this way he gets a ▽. He removes the phlegm from this in a steam bath and he is left with an ♋ which looks white in colour. He uses this to maintain good health, a spoonful at a time. It

101

destroys all impurities in the stomach and keeps it in good health.

Theophrastus's best method: He dissolved ☿ in aqua fortis and precipitated it with ⊖. He sublimated it and ended up with a white sublimate. This he sublimated 7 times. He ground it finely and put it in a flask. He poured half a measure of S.V. over it for every one pound of ♀. This S.V. had been distilled from Floribus Sal Arm. per Retort. He left the sealed flask in a warm place for 14 days and nights. Then he filtered it and distilled it in a steam bath until it became an ♾. He poured the S.V. which he had boiled off over the ☉ which remained in the filter. Then he repeated the process again until the S.V. had extracted everything. He put all the extract in a phial and his elixir was ready. Then he took 1 Lot of ☉, dissolved it in 6 Lots of spirit of ⊖, gradually added it to the elixir in the phial. You must use 8 Lots of ☿ for every 1 Lot of ☉, then he sealed it and put it in a steam bath for eight weeks and he found a red ♾. While it was warm he added a drop to one pound of ☿ and it became the finest ☉. If you add two drops of this to some wine it becomes the very best medicine.

The Author's Tale Continues.

How He Went To Salzburg
and Vienna with the 5 Adepts, and How Finally He
Came Home With Them Again.

Once Herr Schulz said: The air is mistress over the other elements. Without the air nothing can exist. The air exists through the movement of the stars and the heavens and it brings with it all its characteristics and virtues. In short, the air is the focus of our art. For air is to be found in all of nature, and it is the air which gives the philosopher's stone all its strength and qualities. We will learn how this happens. I knew a man who made the lapis which contained true nature. With it he could turn all stones and all metals into the finest ⊙ and cure people from every illness.

He took Sea ⊖, dissolved it in ordinary ▽, filtered and coagulated it. Then he broke it up and melted it. He distilled it, melted it again and always kept the ▽ until he had a fairly large amount. He put this coagulated air, the ▽ in a glass container, buried it in a dung heap of in some earth on which the sun shone all day, and left it to stand for 3 whole months. Then he took the ▽, distilled it per retort with the 2nd. degree of △

and in this way a ▽ evaporates and quite a lot of
living mercury too, which is not well known. He kept
this, took the earth in the retort and poured over
the ▽ which came off with the ☿. He leached it out
in a warm place, filtered and coagulated it and he
found a white ⊖ which was as sweet as sugar and
smelled lovely. Then he took 3 parts of this ☿ and
one part of Miller's ☉ and made a lovely amalgam of
the two. He ground it, poured common ▽ over it and
it turned as black as coal. He poured off the ▽,
poured fresh ▽ over and repeated this again and
again until the ▽ was no longer coloured. Then he
dried it in a gentle △. He took the ⊖ from which
he had completely removed the ☿ and triturated it
with the amalgam until all the ⊖ had disappeared.
He put it in a phial large enough for a third to
remain empty. He sublimated it with Spanish wax and
put it in a Lamp oven with two lights burning, and
treated it until total putrefaction had been
achieved and many colours had appeared. When all the
colours had gone he lit the third light and waited
until the phial had turned as white as milk. When
the whiteness had gone he lit the fourth light and
waited for the phial to become as red as blood. He
did this for every colour in seven weeks, which
makes 21 weeks altogether. When you have 5 Lots of

red ♂, grind it finely with 1 Lot of Miller's ☉ in a little flask. Then place it in a sand bath or over a gentle glow, taking care that the flask does not crack in the heat. Stir it with a clear stick until no more ▽ leaves the vessel. Then take your ♂, which is as yellow as ♀, and put it in a mortar and do the same as before. In the first fermentation it will tinge 10,000 parts, in the second 20,000, and every further time you ferment it and augment it, it will tinge a further 10,000 parts. Then you will see that the tincture is in the ☉, but the materia which unlocks the ☉ is tinged, and that is how the tincture is introduced. This is nature herself. We cure people with this. We do not need the ☉, instead we take half a grain of the red ♂ or philosopher's ♀ and use it to give strength to very old people or to give people fertility. You could also make an old and shriveled tree fertile and green again with it. Bore a hole in the main root and put in as much of your tincture as you would give a human being. Seal the hole with earth, pour ▽ on the root and you will have made a good tree bear fruit twice before the year is out.

Schulz. A miner cemented common ⊖ with three parts potash. Then he leached it out, filtered and coagulated it, melted it all together, dissolved and coagulated it again. He took 3 parts of this mixture and 1 part arsenic alb., and ground them finely together. He put this in wet sand, melted it, coagulated it and poured it into a flask. He put it in the sand again. He repeated this process 8 times and the ninth time he dissolved it with ▽ and coagulated it dry. Then he poured it into a pan which was painted inside with chalk. He added ☿ drop by drop – 1 1/2 pounds of ☿ for every pound of ⊖. For every 1 pound of ♀, 8 Lots ☽ will remain. When it has been flowing for an hour, dissolve the liquid, filter and coagulate it and it will be good again. This is easy to believe because the arsenic is liquid ☽ and the ⊖ joins with it and leads it into the ♀, which is nothing other than a metal. This means that if it is added to ordinary ☽ or ☉ with the correct other ingredients, it will turn itself into whatever you add.

Another One Told by a Silver Miner.

He took 8 or more pounds of common ⊖, mineram ☉is which had passed a ☉ purity test, and mineram ♂, and mineram ♀. Equal amounts of each mineral should be used and 3 parts ⊖, finely ground. The mixture should be moistened with ⊡ and buried in a glass dish very deep in the ground. Leave it there for a long time until the materia has decayed and is all mixed together. Then dissolve it in common ▽, filter and coagulate it, and you have a yellow, almost ☉ coloured ⊖. It has completely eaten all the minerals, so that nothing is left but pure earth, for the ⊖ becomes liquid during the putrefaction process, and attracts the best ♃ from the three minerals which bind the ☿. Then he did the following: He took 2 parts of this ⊖ and 1 part ☿ viva. He did exactly the same as in the first work explained above, and he found the ☿, pars cum parte, half ☉, half ☽. This is how he became rich in a very short time.

Schulz continued his tale: There was a philosopher, a good and pious man who took ordinary ⊖ and dissolved it, filtered and coagulated it until it was see-through, like crystal. Then he took ☿, ground it with the ⊖ for a long time. He believed that the ⊖ would attract the ☿. But this did not occur. Instead, the ⊖ purified the ☿. So he took the ☿, washed it clean and kept it. Then he took the ⊖ with which he had ground the ☿, dissolved it, filtered it and coagulated it again.

107

Then he took calx auri and ground it for a long time with the ⊖. He often left it to stand overnight. It attracted the ⊖ and became a bit damp. It also attacked the ☉ and extracted the ♃, leaving a brown earth behind. He dissolved this ⊖, filtered and coagulated it and it turned red as blood. Then he took his ☿ and added it to this ⊖ and ground them for a long time. In this way he made an amalgam, but some blackness remained which he was very pleased about. Then he put his phial in sand, which was so hot that he could hardly put his hand in it. He heated it until all the blackness was present. Then he waited for the redness. With this red ♂, he cured many illnesses. He also wanted to fix the ☿, but failed. The tincture turned it into glass. Then he fermented it with an eighth part ☉ or ☽ and made ☉ and ☽.

Schulz told us another process and a secret work of nature. You should take as much dew as you can find, which has fallen between May and October and store it in a chip cask made of earth or wood, in a place where there is plenty of fresh air, but where the sun does not shine directly onto it. Let it stand until it is foul and decayed. Then distill it in a water-bath. ▽ steams off and a slimey ⊙ remains in the vessel. You will also get 1 part

☿ viva. Pour the ▽ over the ☉, put it in a warm place so that it can extract well. Then filter and coagulate and you will find a ⊖. When you have 1 Lot or more of this salt, grind it finely with half the amount of subtle gold. Grind it until the ⊖ changes colour. Then grind 2 parts ☿, the same amount as you had ☉. Pour it in and it will fizz and bubble and the ⊖ will attract all the ☉ and leave a brown earth behind. This earth is useless. Put your amalgam into a phial, large enough for three parts to still be empty, put it in the philosopher's oven. The glass must be sealed very well, for the spirit can escape through the tiniest crack. The oven in which the phial is hung must also be sealed so that the warmth cannot escape and the glass is warm at all times. The phial must hang like other glass vessels so that you can see through it. The oven should be put in a light place and light one small light. When it has burned for 5 weeks you will see how the three join, charge colour, decay and rise again. Now your phial will be like the darkest night, now like the clear sunshine surrounded by many colours. Continue with the light until you can see that the nature inside your phial is quiet and calm. Then light the second one and nature will rebel, and the glass will almost fill up completely. The nature will become as white as snow.

Then light the third light until it turns red. When you light the fourth the nature will rebel again and will be redder than before. Make sure you do not move the phial, for that would hinder you greatly. Take 3 parts of this red ♀ and one part of subtly made ☉. Grind until you can see glowing sparks jumping out of the mortar. Then put it in a flask, place the flask in a natural △, making sure that the heat is at the second degree and the glass does not start to glow. Wait until the nature rebels and then it is ready and will tinge 30,000 parts. If you are going to use it as a medicine, however, you must not ferment it but dissolve one grain in a quart of wine. With this you will be able to cure 50 people of all illnesses, as long as you give each only a small glass of it. For if you were to give a whole grain to one person he would die suddenly because the medicine would be too fiery and penetrating. If you want to augment it, ferment it again with ☉ as I told you before.

Schulz told me that there was a man who bought all ores, whether they were ☉, ☽ or ♀. Then he dug a large barrel into the earth so that the top of the barrel was flat on the surface of the earth. In this barrel he ground the ores up finely. Then he poured in all his ⊡ and Stercus, he also put in the ☉ of aqua fortis, old nails and copper plate. He filled

110

the whole barrel with these things. He covered it all carefully with ⊡ and feces. Then he covered it with a board, covered that with earth and left it to stand for a year to make sure the materia opened up and decomposed and that the arsenic and ♄ of the ores were fixed. Then he removed the materia, poured it onto some boards and left it to dry in the sun. Then he melted it in an annealing furnace. He carefully removed all the clinkers so that the King was not robbed of its fire metals. Then he added it to 2 parts ♄, melted and cuppelled it. He was only a distiller, yet he left his children 3 times 100,000 florins.

I told him about a man who came to me who had half-filled a barrel with ⊡. He added old ♀ ground finely, buried it in the earth for half a year, dried it, ground it up, cuppelled it and in this way he made 24 Lots ☽ from 1 pound of ♀

Fauermann told Koller about a chemist who had a refining ♂. He took 12 Lots ☿⟳, made from equal amounts of ⊖ and ♁ and mixed it with 6 Lots ♂ filings. He put this in wet sand, item 12 Lots of Sublimate, and 6 Lots of ♀ filings, item 4 Lots of Sublimate, and 1 Lot of calx auri, and melted everything to an ♋. After a long time he filtered everything, poured the solution of ♂ and ♀ over

111

the ☉, distilled it per alembic, and then applied a fierce △ so that it was sublimated. The water which had evaporated was poured into the condenser, was washed out and then poured over the ☉ and boiled off again. He repeated this process again and again while there was something left to sublimate, then took 8 Lots of ☽, melted it and added 1 quarter of this ♂ and it became the finest ☉.

Schulz said: There was a man who took ♄ of ♀ and poured putrefied and distilled rain water over it. He left it in a sealed vessel in horse dung for 12 weeks. Then he filtered it, coagulated ad liquor, ad hum., and poured as much strong wine vinegar over it as was needed for the liquor to dissolve. Then he filtered it through filter paper, coagulated it again, dissolved it in rain water and repeated this process 12 times until the acidity of the vinegar was gone and the ⊖ was liquid. During the final coagulation it is white, and of the consistency of wax. Take 8 Lots of this, 1 Lot of Miller's ☉, or calx auri made with Regis. Melt it in the cellar and distill it until it is dry. You are left with a yellow ♃. If you add 1 Part of this to 20 parts ☽, will have fine ☉.

Fauermann: Take Sal Armoniack sublimed, 16 Lots common ⊖, 6 Lots lim., ♂, melt it to an ♂♂, distill off the ▽ per alembic, and sublimate it. Wash the condenser out with the water so that it all returns to the ☉. Do this again and again until there is nothing left to sublimate. Then take ☿⚯ and sublimate it 4 times from the ☉ of ⊖. Take 12 Lots of this and add calx auri. Melt it in wet sand and it will extract a. a. Filter the extract and pour it over the fixed Sal Armoniack. Boil off the ▽, heat it fiercely until it melts add one part to 4 parts ☽ and it will be refined.

   Schulz said: Take Ter. cem., and boil it out. Filter it and coagulate until you have a ⊖. When you have boiled out the earth for the third time, dissolve and coagulate it 8 times using rain ▽, then pour a good Seidlein (pint or half Litre) of S.V.R., and stand it in a well-sealed container in a warm place. Leave it until the S.V. is blood red. Then filter and distill it and pour the S.V. back over it. Repeat this process until you have extracted everything except for a little grey earth. This cannot be dissolved. When all the S.V. has been distilled from the extract pour the S.V. back into the flask. Distill it again and some feces will remain again. Repeat this until there are no longer any feces in the filter and until the ⊖ is as white as a lily. Take 4 Lots of this ⊖ and grind it for

12 hours without stopping with a quarter of Miller's ☉ so that it heats up nicely. Sprinkle some distilled rain ▽ over it so that the ⊖ is moistened but not dissolved and then grind it again without stopping for 6 hours. Dry it over a gentle glow and it is potable. Put it in a phial large enough for 3 parts to remain empty and sealed at the top with a blister, place it in horse dung, making sure that the tube is not buried too, otherwise it will steam too much, and leave it to stand until it has turned into an ⚇ which is as red as blood. The feces must be separated from this ⚇ by the secret art. One drop of this ⚇ is given in all circumstances ad Lapidem. (It must stand for about 12 weeks in the dung before it becomes ⚇.) Put it in a phial sealed with a cork and a blister, put it in a Lamp oven, light one light until it turns black, then light the second light until it turns white, then the third when the whiteness has gone and it will become a red ♂. If you give a sick man half a grain of this ♂, he will recover. It is the best ♄ in the world. It can make your grey hairs fall out and make new hair grow. Leave it to stand for 6 weeks, then light the fourth light as soon as it turns red. 4 lights must burn for 6 weeks. Then it is ready except for the fermenting. Take 4 Lots

of this ♂, 1 Lot Miller's ☉ and grind it until all the ☉ has been consumed by the powder. Then put it in a flask in a sand cupel, stuff paper into the opening to seal it and heat it at the third degree of heat in a fixing furnace for 12 hours. Remove it and you will be able to tinge 20,000 parts according to our art. If you wish to augment it take 8 Lots of the ⊖ and one Lot of the tincture and follow the above instructions.

N.B. Earth is good at all times, but it is at its best in the middle of May.

Another one: Dissolve 1 pound of ☿ in aqua fortis made of equal parts of ⊕ and ♅. Precipitate it with common ⊖, wash out the precipitate and take ⊖ as I taught you and S.V. Take 16 Lots of this and grind 2 Lots of Miller's ☉ with it. Grind until you can no longer see any of the ☉. Then grind 16 Lots of ☿ ⏚ with it. Again, grind until you can't see it any more. Put all this in a phial, sealed with a blister, stand it in horse dung until it turns into an ♋ (as I have often described). Then filter the ♋, put it in a little flask, place the flask in ashes (it would be safer in a Lamp oven) and let it change through the three colours as I described before. The third colour is red. Ferment this red ♂ in an open △ in a little flask or well-

115

sealed phial with a sixth part ☉ and you will tinge
50,000 parts. You will be able to cure all
illnesses, prolong life and similar things. All this
must take place before fermentation with the
exception of transmutation.

He continued: There is another earth which is,
however, not as good as this one, but the stone can
still be made with it in the following way: Dig a
hole in a field two spans deep. Fill it with human
feces and urine and cover it over with earth. Leave
it for a whole year. Then remove the top span of
earth which is useless because the volatile ⊖ is on
top where it can attract the volatile ☿. Dig for a
further 2 or 3 times, filter and coagulate it and
you will find a red ♂. It isn't as good as the one
I told you about before, because it has a lot of
common ⊖ in it. It will only tinge 20,000 parts at
the most.

There was a man who put vine branches (of the
fat sort) into a phial, sealed with a bung and a
blister and stood it in horse dung for 6 weeks. Then
he filtered it per Retortam. He was left with an
♑ and a ⊖ volatile, rectified. Now distill the
spirit through a condenser and extract the phlegm.
Then pour the ♑ and the phlegm over beech ashes and
rectify it per Retortam. You will be left with a red
♑. Separate this in a funnel, dissolve 8 Lots of ☽

116

in 16 Lots of aqua fortis precipitated with ♀ plate, wash it out, take 8 Lots of this calx, 24 Lots of ☿ and make them into an amalgam triturando. Wash it with ▽ and put it in a sugar glass and pour 16 Lots of your ♋ over it. Leave it in a cellar for 8 weeks and many little twigs will grow out of it. Cut them off with a sharp pair of cutters. Cupel it and you will have fine ☽. Add the same amount of ☿ and 3 or 4 drops of ♋ all the time and you will have made a particular. The universal work is made as follows: He dissolved ♀, in aqua fortis and precipitated it with common ⊖ and 1 pound of ♀, in 2 pounds aqua fortis and then burned the vine branches to ashes and extracted the ⊖. Then he took 2 parts of this and 1 part of the above ♀, ground finely, put them in a flask and poured the ♋ from before over it. He used twice as much oil (in weight) as he had materia. Then he distilled it per alembic, first very gently, then fiercely, and the ☿ began to sublimate. But it was black in colour. Then he took the ♋ which had been boiled off and washed the condenser with it. Then he added it ad Lunam again and repeated the Process. He kept repeating it while there was something to sublimate, otherwise it will flow in the flask like pitch. The heat of the △

117

should be constant. Then add to 8 Lots of the materia, 1 Lot of ⊖ calx, ground finely and let it stand for 24 hours in flux. Then the ☉ must be added as soon as it begins to smelt. It will look yellow in colour. At this point it is left to cool down. It will tinge 12 parts, all metals which are as liquid as ♄ and ♃, ☿, ♀ and ☽.

On another occasion Koller said to Schulz: I think it is best to use metals in our work. All things come from God. The seed which I plant comes from me in the same way. If I bring ☉ to putrefaction and bury it in earth it will multiply like wheat and corn. My menstruum is nothing other than ☿ viva which decomposes the ☉ ex fundamento and brings it to putrefaction. For the more often I plant the seed, the more often I can reap the fruit. This is why I prefer the metals.

Schulz: I do too, although I am not saying that there is a tincture other than the ☉. But I am saying that there are many different menstrua which can dissolve it. If I planted wheat in bad and hard ground, it would grow, but only badly and only a few of the seeds would sprout. But if I were to plant it on good ground it would bear me one hundred times as much fruit. This is what our art consists of. If I have a good menstruum which dissolves the ☉ ex

Fundamento I will be able to tinge a great deal. But
if I have a bad menstruum I will tinge far less.
There are many menstrua but only a few good ones.
I prefer the ones I have told you about. I will not
say that it is not in the ☿, neither will I say it
is, nor will I say it is not in the ♄ of the metals.
It is true that there is a good menstruum in them,
but they are not as good as ⊖. Of course all of
these must be in ⊖, but it is better if I can take
the ⊖ from nature, than if I try to make a ⊖ from
☿ or ♄, for the true menstruum really is nothing
other than ☽. How do we know this? Why can an
♋ ♄ous or an aqua fortis not dissolve ☉ ? Because
the aqua fortis is hardened with ♋ of ♀, and its ♄
is similar to the ☉. Why does the ♋ of ♄ not
operate? Because it is also a friend of the ☉, that
is why it will not do it any harm. But spirit of ⊖
is an enemy of ☉ and that is why it tears it open
and destroys it. So it is easy to understand why
nothing will dissolve the ☉ other than a spirit of
⊖, because the ⊖ is the true menstruum which opens
the ☉, makes it subtle and introduces everything
into the gold. For when the ⊖, goes through
putrefaction it enters ☿ and binds it as I have
taught you before. It also unlocks the ☉ ex

119

Fundamento. But not only does it unlock it, it also enters the raw metal and fixes it. But it transmutes the fermentum, Mercurius can do it too if he is made into a ⊖, but without this it is difficult. Koller said: Dear brother, I was able to tinge 10,000 parts without turning ☿ into a ⊖. I joined it with ☉, ground it until the whole corpus of ☉ had become ☿, and they go to putrefaction mixed together.

Shultz: Do you know what putrefaction is? What happens during the decaying process? Everything separates. Good and evil separate. Which colour are you left with after putrefaction? Koller said: I am left with white crystals like ①. Schulz: But what do you find in Fundo? I find brown earth and underneath this a fixed ferment. Schulz: Do you need this earth and ferment? Koller: I don't need it and I do not separate it until it has turned red. But when it is red I take out the phial and the red ♂ is loose on top of the brown materia. I remove this and at the bottom is a fixed brown earth together with the ☉ I added. Schulz: What do you think is the significance of the earth and the red ♂ on top? Koller: It means that the garbage has separated from the ☿, and the fixed putrefaction which was not ☿enough has sunk to the bottom. The red I take to be the ♃, which is true. Schulz: That is correct, it is

120

the true ♃, but what does it consist of? It does not consist of ☉ or ☿, but of ⊖. This forces the ☿ as it does the ☉. It separates during putrefaction from its ⊖. The ⊖ rises because it is light, but the ♃ stays at the bottom because it is heavy and fixed. For is your red ♂ not a Soluble, which will dissolve in any ▽? Koller answered: Yes, it dissolves in wine as well as in ▽, therefore it is not a ⊖. If you put a ☿ ⎯ into water, the water will attract the opened ⊖, but lets the earth sink to the bottom. It is obvious, therefore, that the ⊖ must be drawn out of everything. Your work is useless if the chaos is still together, for the lapis itself, when it has become white and fixed, is nothing other than a ⊖. To make a menstruum of a ⊖ to decompose the ☉ you only need to purify it until it becomes a liquid. I said to Schulz: Dear brother, you say that there is nothing of any value to our art other than this ⊖, the nature of all things, which I recognize and know to be true. But where does this ⊖ come from? Schulz said: It rules the air, and is made from air and △[13]. Koller said: But where does it get the air from? Schulz said: Go and ask God how he created nature! God's work is not to

---

[13] This is the symbol that Hans used. -pnw

121

be understood. We know what it consists of, but only God knows how he made it. His omnipotence is not to be explained and there is no one who will find the answer in this world.

Monte Schider said to Schulz: Dear brother, would it be possible to bind ☿ without using a tincture? Shultz answered: This can easily be tested. If you were to introduce ♁ it would surrender itself immediately and no longer float in the wind. To make it solid you must stop it from flying. Ruesenstein said: Dear brother, I fix it to ☽ without using a tincture, and to ☉. But it was extremely difficult. It is not easily done. Shultz: It is true that it is difficult. Whoever wishes to prepare ☿ must ferment it, but it will not be fixed with a fixed metal. The metals must be decomposed. Only then can you introduce the ☿ and only then can it be fixed. Ruesenstein said: This is what I meant. Listen to me while I tell you how I do it:

I take one pound each of ♄ and ♂ and make a chaos from them. I pour it through 2 parts ♄ and melt the ▽ and impregrate it with as much ♂ as possible. Then I amalgamate it with ☿, triturate it until the ☿ looks tender and I cannot see any more ▽ in it. Then I wash it out and dissolve 4 Lots of

☽ in aqua fortis precipitated with ⊖. I take 8 Lots of this Luna and 1 pound of the ☿ from before. This is ground in an iron mortar until a soft amalgam has been made. Then I grind it for a further 24 hours, day and night, in a triturating mill in a warm place. Then I dry it and put 8 or 16 Lots in a phial and place it in the sort of oven where △ can be applied from above and below (as shown in a compendium). I apply △ for 8 days, at the second degree of heat from below. Then I apply the third degree of heat from above for 8 days, then the third degree of heat from below for another 8 days. Then I apply the third degree from above and below at the same time for 4 days and nights. I am left with a blood red ♂, as well as a metal on the bottom. This metal should be ground and mixed with the ♂. I put it in a warm mortar with twice as much ☿ aati, ut ante, then I ground it until I had an amalgam and then carried out the rest of the work as before. I did this three times. The first time I tried it in a cupel. Half of the ☿ remained and in the core was about 4 Lots of ☉. The next time about 3 parts remained in the cupel, and it was pars cum parte. The third time the cupel was red. I did this work with my own hands and I know it to be true. It is true that it is better to work with water-mills than human hands, but what follows can be done with the human hand. That is:

I took 1 pound of common ♃ and melted in a pot. I added 1 pound of ☿ squeezed through leather, stirring vigorously all the time until it looked like pitch. Then I ground it to a ♂⃛, lit it to burn off the ♃, and I was left with a coarse ♂⃛. I took 1 part of this, 3 parts potash and 1 part iron scale and distilled it per Retort. Then I got some fresh ♀, I prepared the ☽ in aqua fortis as before. The only difference is that the ☿ is prepared in a different way. This can be done by hand. The former is horse's work; so much grinding in order to animate the ☿ with the ♁♂ to make ♃ic ferment. Shultz said: I like your method, but I have another easier one.

I take ♀ precipitated with aqua fortis and washed. I imbibe it every day with the following ♋ of ☿ and it fixes itself. This is how I make the ♋: I dissolve ♀ in aqua fortis and evaporate it ad liquorem. Then I add common ▽ precipitated with common ⊖. Then I wash out the precipitate and dry it. Then I pour distilled vinegar over it and extract it. I keep pouring vinegar over and then extracting it. I evaporate the vinegar, then pour ▽

over it 3 or 4 times and evaporate it again until all the acidity has gone. Then I put it back in a flask and pour over 1 measure of S.V.R., cum ⊖ of ♀. I extract it, distill the S.V. off per Balneum and I am left with the ☍ of ♀ with which to imbibe the ☿ precipitate. It must be imbibed 8 times and it will tinge the ☿ with itself. Then take 6 Lots of fine ☽, melt it in a crucible and gradually add the ☿ bit by bit. Leave it melting for half an hour, then pour it out and you will have ☽ rich in gold. It will give a third part of ☉. Everything exists in the ⊖, for as soon as a metal is opened it is a Soluble or ⊖, and it can be easily extracted and turned into an oil. Briefly speaking, when a metal has been opened up the volatile parts of it are no longer bound by the fixed parts, and the volatile parts fly away and the fixed parts remain behind. If the fixed parts are made volatile and added to the volatile parts, they join immediately and you will enjoy the fruit of their union. The ⊖ is the most important thing if you want to change a metal. Use it for your profit.

Fauermann: Take 1 pound of ♂ filings, 2 pounds of ♓ sublimated. Turn it into an ☍ in the cellar. Pour ordinary ▽ over it and apply sublimating △.

The ♂ will be sublimated yellow. I washed out the condenser with the ▽ and then returned it to the flask. I sublimated it 8 or 9 times and the Sal ammoniac remained fixed with the ♂. It melted together to form a red stone. I dissolved this with distilled or filtered aqua fortis, precipitated with vinegar and yellow ♃ sank to the bottom. I washed it out and gently dried it. Then I took ☽ cornua, item ☿ with ⊖ ut ante precipitated. I dissolved 1 pound of the ☿ with vinegar for 12 hours, and when everything had been extracted I precipitated it again with ⊖. I poured it over the precipitated S.V.R. with ⊖ of ♨ 3 times. Then I dissolved it and filtered it again, extracted the S.V, and I was left with an ☍. Then I took 8 Lots of my ☽, 5 Lots of my ♃ and ♂, and ground them together. I then poured 2 Lots of warm ☍ of ☿, over it to make an ☍. When it is cold it turns into a ⊖. Stir it and leave it in a place so hot that your hand cannot bear the heat for 24 hours. At the end of this time it will be dry. Dampen it again with 2 Lots ☍. Do this three times. Then take Luna cum ♃ and cement and seal with luto in a pan. Then the ♃ melts with the ☽, and the ☍ consumes them together. Then a smelting △ should be applied and the ☽ sinks to the bottom in a compact form, red in colour. Then it should be cupelled and it will remain red for ever more.

Shultz said: Theophrastus once performed the following work: He dissolved ☿ in aqua fortis and boiled it off per Retort so that he was left with a precipitate. He also made a precipitate from ☿ with aqua fortis and common ⊖, as you already know. He extracted it with distilled vinegar as I told you before, then he precipitated it again with ⊖ and extracted it again with vinegar. He repeated this process until the vinegar wouldn't let the ☿ go, and no white substance sank to the bottom any longer. Then he boiled off the vinegar and was left with an ⚇, or rather a spirit of ☿ which can unlock all metals. This he poured over his first red precipitate which must be washed out. It began to boil. When this happened he found a coal black fatty ♂, of the consistency of wax. He filled a phial with this and placed it in a fixing furnace in the sand. He proceeded to heat it with △ in the manner described in books. Within 24 hours he had achieved the black colour, in a further 24 hours it was white, and finally it turned red too. He took 20 Lots of this red ♃ and 1 Lot of gold and put them in a very warm place so that the gold decomposed. Then he took 1 Lot of this, added it to 3 pounds of ☿

while heating it, so that it started to drink.

Within a quarter of an hour it had turned into ☉.

Why did he only use 1 Part of ☉ to 20 parts? This

is because the ☿ is a ◑ solis and contains the

nature of ☉ within it. Theophrastus only worked

with ☿, but he talks about it in so many different

ways that it is impossible to understand him

sometimes. He calls it life, a beautiful virgin, the

evil thief in the heart; he also calls it an eagle.

He had three methods of working, but all three used

☿. The first was a ☿⚊, as I will explain below,

with ☍ ♁ is per campanam; the second is ☿⚌ which

we were thinking about before when we remembered

that it is fixed using ☉. The third was an ☍ of ☿.

I have already mentioned the latter, but I had

better tell you the whole thing.

He dissolved ☿ in precipitated aqua fortis and

washed it out with common ⊖. Then he sublimated it

and was left with a crystalline sublimate. He ground

this sublimate finely and extracted it with

distilled vinegar until almost everything had been

extracted. Then he boiled it until it was dry. He

poured ▽ over the remains and smoked it off again

to get rid of the acidity of the vinegar. Then he

ground it again and poured S.V.R. over it and left

it to stand in a warm place. He did this several times until he had extracted almost all the S.V. Then he filtered the S.V., distilled it in a Steam Bath and he was left with an ♋ which will dissolve all things; ☉ and ☽ and things like itself, namely ♀. Then he took 8 Lots of this ♀ ♋, and added 1 Lot of calx auri, placed it in a phial in a Lamp oven and left it there until the ♋ had joined with the ☉ and was blood-red in colour. Then he removed it when it was cold and it was hard. But if it was warmed it turned into an ♋ and could penetrate metals and unlock ☉ ex Fundamento. If you do this ♀ becomes fixed and loses its wings and will ever remain fixed in a ♃.

Fauermann said: ♀ penetrates and opens everything. If a metal is unlocked with it, it cannot be lost. If ☉ is dissolved to a ▽, until you think it cannot be reduced, just add a few drops of ♀ to it and you will have an amalgam. But if it is dissolved with something from which it appears that it cannot be separated, it will make a wonderful tincture. Whoever can work with ♀ properly is a master of the art.

Schulz: This does not only apply to ♀. There are others which, if their ☉ is purified and made wax like, will attack ☉, unlock it and penetrate it and will not be precipitated from it. ☿ is a ⊖, which has no ♁. That is why it achieves the ♁ of life. Because ☿ is a ⊖, and because ⊖ unlocks everything, I call it the master of metals.

Fauermann continued: I have heard of many ☿s. There is ☿ of ♄, of ☉, of ☽ and ♂. Which is the best? I don't consider one to be better than another. For whatever is a liquid ☿, is a ⊖, that is a dead corpus. If it had a dignity, that is an amalgam of those metals of a ♁ in it, it would not be a liquid but a coagulate. But while it is liquid it contains nothing other than itself. And that self is a ⊖. But how, dear brother Schulz, is the ☿animated so that it takes an animation upon itself so that it remains constant in the △ more easily and can be more easily fixed? Schulz said: It is not animated but rather purified. At the same time it unlocks the metal with which you animate it. But it can easily be separated from it again. And this is how ☉ is made from ☿. It must be purified for it is almost less pure than that which is brought up from the mine and contains much raw metal which must be removed. But how? Through the ♁ or ⊖; but which is the best way to purify the ☿? The best way is as follows: I dissolve the ☿ in aqua fortis with common ⊖ precipitate. Then I revive it and it is as pure as possible. But how is this done? Like this: The ☿ should be dissolved with an equal part of its weight, add as much ⊖ as can be precipitated. Mix the precipitated ♂ with potash viva, added to half as much rye or wheat flour, distilled per retort. At the beginning a red spirit rises from it, then an ☍. Then the living, clear ☿ leaves it. This is so greedy that it is capable of running 3 spans towards the ☉ and dissolving it in a split second. This is the best way of purifying the ☿. Then, if I want to purify and clear ☽ which is contaminated with other metals like ♀, I dissolve it in aqua fortis, filter it and precipitate it with common ⊖ and the ☽ sinks to the bottom like milk. But the other metal remains in the ▽. This is the best way of purifying ☽. ☿ has the dignity and crudity of other metals in it which it will shed like the Luna in aqua fortis. If you do this to purified ☿ you will lose the fourth part, for the impurities and foreign bodies stay in the ▽ and cannot be precipitated. Take some of your ▽ and heat it until it smokes. You will see the filth in it which used to be in the ☿. It is as black as ink. If you put it over a glowing fire it

131

disappears like lightning before it ever gets warm.
This is the poison which stops the ☿ from being
fixed. If it is purified as I have just described,
however, it is good for everything, for metals and
metallic things. But if I revive it as I explained
before, I am left with a spirit and an ♋ which I do
not throw away. I know a good use for this too. For
the spirit extracts the animated ☉ and lets the
corpus of ☉ sink like white ashes to the bottom.
This can easily be reduced with a 🜂 of ♂. For if I
distill my spirit from the animation per Retort, I
can use it further with the anima to turn as much ☽
into ☉ as there was ☉, from which the anima was
extracted. It can be used for all wounds to the
human body - new or old. It takes away the old
flesh, pulls the nerves and veins together and
promotes the regrowth of flesh. But more can learned
from this: Take your ☿ rectified and put it in a
retort. Distill it. When it has been distilled you
are left with a small corpus, a red precipitate.
Keep this. Then distill your ☿ again. Repeat this
until almost half of the ☿ has become precipitate.
Then take 1 part of your precipitate to 2 parts of
the ☿ viva and grind them and make an amalgam. Fill
a phial with it and put it in a sand bath. Apply △
in the normal way as I taught you before. In this

way the amalgam goes through putrefaction. When this is over it will have become a white ⊖ which will soon become earth. Ferment it with an eighth part ☉ in a pan or glass as I told you before. Augment it with 2 parts of the same rectified ☿ and apply △ as I taught you. Do this three times until you can see a red colour. It will tinge 20,000 parts.

Shultz once said while we were on our journey: Dear brothers, do you know anything else about ☿ ? Koller said: I know something which also results in an Augmentum, but the purification is very difficult and liable to fail: I dissolved 1 pound of ☿ in 2 pounds of aqua fortis, boiled it off again and poured the aqua fortis back over it. I did this 3 times and was left with a lovely red precipitate: I revived this with potash and iron filings and dissolved it again in aqua fortis. When I had done this 3 times I didn't do it again a fourth time. I put it in a phial and it was precipitated without fermenting to a ☿ial and it refined ☽ into ☉. The reason why the colour of this ☿ is raised is that the aqua fortis contains ♀ or ♀ from ♎ which penetrates the ☿. When the ☿ is revived it takes the subtle ♀ with it and is animated, so that the aqua fortis is not powerful enough to dissolve it any more.

Ruesenstein said: I too found something in the ancient writings. I did the following: I took 1 pound of ☿, 3 pounds of ♃ common, and melted the ♃ and lit and distilled it with 3 parts potash and 1 part iron filings just as I told you before. When I distilled it per Retort, a yellow ☿ evaporated. At first I was shocked for I thought the ☿ would not evaporate. Then, when I had applied a fierce △ for a long time, as one does when one wishes to distill ☿, I did not see any ☿ for a long time. I was angry about this and lit △s all over the oven. Then gradually the ☿ steamed off. I took 8 Lots of this ☿ and mixed it with 1 Lot of calx auri to make an amalgam. I ground it with a triturating mill in an iron mortar until it was as black as coal. The ▽ was purified and distilled rain ▽, either collected during a thunder-storm, or melted from snow, or collected from a puddle near a mountain where nothing other than rainwater can be found. I distilled the black ▽ and was left with a black ♂, which I put to one side. I poured the ▽ I had boiled off back into the mortar and ground it as before until it turned black again. This has to be done until you cannot see any ☉ or ☿ in the mortar. I gently dried this black ♂ and when it was

134

completely dry, I took twice the weight of the same ☿ rectified, and amalgamated it with the black ♂. I had to make the amalgam into a black ♂ again. This took 8 weeks, I did this 3 times and the third time I put 2, 3, or 4 Lots of this ♂ in a phial in the sand and heated it with a fairly strong △ (at almost the third degree of heat). A little of the ☿ viva could be seen in the tube. It circulated for 14 days and nights. When it stopped circulating I had to increase the △ fully to the 3rd. degree, and the phial stood for 14 days and nights in the sand, glowing brown. Then I removed it and found a blood red ♂. I could cupel it as much as I wanted and was left with the finest ☉. This is augmented as follows: Add 2 parts of the black ♂ to one part of the red ♂. Grind them and put them in the △ as before. Half of the mixture can be separated by distilling and that in turn can be augmented.

Shultz said: A beautiful amalgama is hidden in ☿ and this is why it only ever precipitates itself in ♃ to a red colour, whether it is fermented with ☽, ☉, ♀ or ♂. The highest redness in nature is hidden in it, but you cannot see it because a volatile arsenic hides it. But if it is burned with ♃ or aqua fortis, its ♃ will not burn, but its arsenic will. As soon as this is gone it shows a yellow colour and is not as clear as it was before. This is why there are many people who make the ☿, ☉is, or ♂, ♀, ♄ and ♂ in the belief that this ☿ does not contain as much poisonous arsenic. But it is all the same; such a ☿ has as much as common mercury, although a metal has no poison when all three principles are together. But if they are separated, then the mercury is just as poisonous and must be prepared.

There was a man who melted 3 pounds of common ♁ in an iron pan and added 1 pound of ☿, pressed through leather. Then he stirred it until the materia was dry and hard. Then he broke it up, mixed 1 pound of common ⊖ with it and sublimated it. He was left with a green sublimate. This he mixed with 3 parts ♃ and revived it. Eight times he sublimated and revived it. The ninth time he sublimated it but did not revive it. Instead he took 8 or 10 Lots of it at the most and mixed it with 1 Lot of calx auri. He ground the mixture together, poured it into a well-sealed flask and sublimated it with the fiercest △, because it does not sublimate easily. It sublimates to a yellow colour, like ☉, and before it is sublimated, the condenser turns as black as soot. This blackness signifies the uniting of the ☉ and ☿. You also know that a sublimatorium is happening. Now you do not need a condenser, instead you should use a little flask above and below, joined in the middle. So when it has sublimated upwards, the following day you can turn it over. Repeat this until the ☉ and ☿ stay together and become a red, see-through glass. Then he ground this stone very finely and added a ninth part ground Miller's ☉ or calx auri to it. He put this in a sealed flask into a glowing fire until the flask began to glow. (But the fire shouldn't be so hot that the glass melts). Then the ☉, joined with the stone and turned yellow. It was hard, but full of holes as a sponge. He used this to tinge 20 or 30 Parts ☽ into ☉. But it is too weak to be used for other metals. But if he had wanted to tinge all metals he would have done the following: He ground the stone, poured over 3 fingers of S.V.R. and extracted it. He distilled off the S. V. and poured it over the ☉ again and repeated this until the ☉ was no longer red and nothing more would be extracted. He distilled off all the red extract at once. If he had used 5 or 6 Lots, he got 2 Lots from 5 Lots. That which was left is useless; it is a dead Terra ☉is and ☿. For the spirit of wine has drawn

out the pure anima. When he had 3 Lots of this
extracted anima he added 1 1/2 quarts of Miller's
☉ which he had dissolved, and placed it in a phial
in a fierce heat. The liquor coagulated to a
sulphur-coloured ♂ with which he could tinge 5 to 6
hundred parts in all metals.

Fauermann: The ☿ takes the anima of ♁ with
it. This method is good: It is dissolved in aqua
fortis, washed out with a precipitate of ⊖ and not
renewed but sublimated 8, 9, or 10 times and it will
solidify to a yellow glass. The ninth part of it is
☉. It is used as explained before.

Ruesenstein: The ☿ is the thing of all things;
a miracle of the world. A metal like this can join
with all others - firstly with itself, but then it
also loves ☉, ☽, ♄ and all metals. If it is
purified, it will purify the other metals. It turns
into whatever it is added to. Coagulate it with that
which it is fermented in, then you will find an
anima in the fixed metal with which it joins and
fixes.

Schulz: This is why the ancients said of ☿ that
it represents the whole art. They didn't mean common
☿, they hardly knew that - they meant the
☿ of the air, of the ▽, and of the earth. For in

all ▽, in all earth and in all air ☿ is to be found. But it is not metallic. It is a ⊖, a solvent and a solution, to which even our ☿ must be brought before it can tinge. For what is ☿ ♎ ? It is a ⊖ of the air, it leaves its earth behind. The ancients took ⊖ from the sea, ⊖ from the air and from the earth. These three all come from the same place. One is as good as the other. Whatever is in the △ is in the ▽. Whatever is in the ▽ is in the ▽̄. How do they get here? Through the air. For the earth and the ▽ are also hidden in the air. The ancients saw their ☿, their ⊖ hidden there. They prepared it by dissolving and coagulating alone, so that it could penetrate and become spiritual and sharp. Through this, evil can be made to leave the human body. It will warm a man up again and help him stay healthy for a long time.

Schulz said: Collerus makes a ♋ of ♀ using aqua fortis. He extracts the ⊖ with vinegar. He boils off the vinegar with ♉. Then he adds little pieces of wood and a little ♀ plate and puts it in the cellar. This is how the ⊖ crystallizes. This ⊖ is as solid as ☉. It cannot be distilled, nor can it be burned. But it turns to glass in the △. He ferments this noble jewel with a tenth part of ☉.

He puts it in a sand bath and applies the second degree of △ per ascensum, until the materia flows in the phial like pitch. Then he applies the third degree until it turns red. Then he multiplies it with 2 parts ⊖ of ☿. He does this 3 or 4 times and the medicine is ready. He ferments it with all metals with a tenth part Miller's ☉ and tinges 10 to 12 thousand parts. Note that this is the most secret and noble work in our art; for it consists in ⊖, in common salt as well as the metal ⊖, but it is quicker if the metal ⊖ is used rather than common ⊖ without the ⊖ hois; the metal salt only needs several dissolvings and coagulations and it is ready to be melted. This is because it has already been putrefied. But common ⊖ must be dissolved about 100 times in order to putrefy and purify it. You cannot see this putrefaction, you can only smell it. For there is a corrosive spirit hiding in it, and only when this has been extracted is it pure. When it is pure it is of the consistency of wax and only needs a ferment, ☽ or ☉.

Another man melted common ⊖, dissolved, filtered and coagulated it about 50 times and then fermented it with a bagatelle[14] of ☽ and in half an hour he had tinged ☿ into the best ☽. From this we

---

[14] The text notes that this is a weight. I think the author means "small bag." -pnw

can learn that the true menstruum of our art is nothing other than ⊖; for without ⊖ no metal can be unlocked. Some salts are better, or at least take less time than others, but all of them come from the same thing. The above author believes that ⊖ of ☽ could be used if you do not dare make the ⊖ of ☿. He describes this in the following terms: Take one part ☽ dissolved *ad oleitatem* in 3 parts aqua fortis, and then abstracted. Pour 3 times the weight of vinegar over it, put it in a cellar, and it will crystallize; pour 1 measure of S.V.R. over 3 or 4 Lots of this. The spirit of wine must be so strong that if you were to let a drop fall 4 spans above the earth, nothing would actually reach the earth. Extract it until you can see a white chalk on the bottom. Then distill the S.V. off until you have made an oil. Now you have made ⚯ of ☽, which you can use instead of the ⚯ or ⊖ of ☿. He also teaches that the tincture is made of ⊖, ♁ and ☿. But you must not use ☉, ♁ and ☿ because ⊖ and ☿ stay together and when it is fermented the ♁ will have the primum principum. This is almost as much as the ☿ and the first and the last thing can be made from it, but the third is added during fixing.

Fauermann: An ancient author used ⊖ of ☿ in the following way: He dissolved ☿ in 2 parts aqua fortis, made of ⊖ and ⦶ boiled it dry. He put the ⊙ in a little pan, heated it and then cooled it with sharp distilled vinegar. He filtered the vinegar and dried the remaining ☿. Then he poured the extracted aqua fortis back over it, boiled it off and then rinsed it with vinegar again. He repeated this again and again until all the ☿ had gone. Then he distilled off the vinegar and was left with a white ⊖, which is a fixed liquid. He dissolved this ⊖ again in fresh vinegar, then filtered it and poured off the vinegar and he was left with ⊖ of ☿. He put it in a phial in the sand and heated it until it glowed brown and until he could see a red colour at the bottom of the phial. He ground up this red stone and added it to a further 2 parts ⊖ of ☿ and repeated the process a third time. He fermented it with a tenth part of Miller's ⊙ and tinged 15,000 parts. He also has a Particular using this: He let one pound of flores of Sal Ammoniack and 16 Lots of ♂ calcined red; melt in a cellar for 8 weeks. Then he poured rainwater over them to dissolve them and then boiled off the ▽. He applied a sublimating △. He washed the sublimate with the ∇ he had boiled off in a flask. He did this 7 times and was left with ♂ and Sal Armoniack at the bottom. To this he added the same amount of aqua fortis which extracts the ♃ ☉is from the ♂ - first it is red but then it turns green. He filtered and coagulated this, dissolved it in rainwater and precipitated it with distilled vinegar and the ♃ ☉ sank to the bottom. Then he dissolved 16 Lots of ☿ in 1 pound of aqua fortis, boiled off the aqua fortis and poured distilled vinegar over the precipitate. He extracted it, precipitated it with common ⊖ and washed it out. Then he dissolved 1 pound of ☽ in 2 pounds aqua fortis and distilled it to an oil. Then he dissolved 16 Lots of ♀ in aqua fortis. He added this solution to the solution of ☽, precipitated it with common ⊖, and then took 16 Lots of this ☽, 4 Lots of ♃ of ♂, 6 Lots of ☿ precipitate and ground them all together and put the mixture in a smelting pan sealed with luto. He left it to cement for three hours in a circulating △. When it had melted he poured it out. He smelted 3 parts ☽ and dissolved it in aqua fortis. He found calx auri at the bottom, which made 3 parts, and remains above 1 part ☽. Schulz said: People say that when the ☿ is

143

circulating and precipitating it turns the most beautiful ☉ colour; ♂ also has a noble △☉is. If these two are fixed and added to the ☽, you cannot fail.

Gualdus makes a Regulus like this: He takes 1 pound of crude ♀, and the same amount of ♂. He makes a chaos of the two in a very fierce △ without adding anything else. He pours 1 part of this chaos through 2 parts crude ♄, as if he were pouring it through ☉. Thus the pure amalgama ♂ and ♄ falls through and he is left with a starry regulus. He pours this through 2 parts ♄ again. He does this 7 times. This is how he clarifies it at the end: He takes 8 Lots ♅, 8 Lots common ♃ and 4 lots crude ♇, 4 Lots ①, 3 Lots common ⊖, and grinds it all up in a regulus. He fills little containers with it - in each one just the amount which would cover a knife-point - then he puts one after the other into a glowing smelting crucible and leaves it to melt for 1 1/2 hours. Then he pours it out. He keeps the clinkers to make a particular, he melts the King and adds a little boiled ⊖ to it. Then he puts it in some ▽ and it becomes purer. He dries it in the air, melts it and adds 1 Lot of good quality ☉:

Then he makes an amalgam from 2 to 3 parts ☿ prepared with 🜍. He grinds it for four days in fresh 🜄. He puts it in a phial sealed with a bung so that no dust can get in. Then he puts it in the sand, covering the bowl of the phial with ashes and applies the second degree of 🜂 until it turns black. Note that the ☿ should not circulate, it must not be separated from the chaos. He continues to heat it and crystals are formed. They should not melt otherwise it will become a regulus again. Then it must be coagulated again and the ☉ will be precipitated again. The subtle spirit turns black again, and then white and red, then nothing more is precipitated. Then ferment it immediately with an eighth or ninth part.

Fauermann told us about a man who made the Lapis from ☿ and ☉: He dissolved the ☿ in aqua fortis, distilled it and sublimated it with ♁ and ⊖, revived it with potash, amalgamated it with ☉, and triturated it until it turned into a black powder; then he dried it gently, put it in a phial and proceeded in the usual way.

Theophrastus dissolved ☿ in aqua fortis, precipitated it, washed it with common ⊖ and sublimed it with 2 parts common ♆. He took 12 Lots of this and 2 Lots of subtle ☽. He sublimated both

145

together until it would sublimate no more. When it was fixed he took 10 Lots of it and cemented it with 2 Lots of Miller's ☉ or calx auri for 2 hours in a glass flask. Then he tinged 6 or 7 thousand parts. The subtle ☽ is Luna cornua.

Theophrastus's universal work: He took some ☿ and dissolved it in ♋ of ♃ per camp. Then he boiled off the ♋ per Retort. He washed the precipitate and dissolved it again. He did this 7 times. Then he took 8 Lots of it, cemented it for 2 hours, gently for most of the time and a little more fiercely at the end, and the precipitate became a liquid. He took the remaining 24 Lots of ☿ ⚌ and ground it and mixed it with 3 times the quantity of wheat flour. He distilled it per Retort. First a spirit evaporates, which can be used for many feverish illnesses. Secondly a black ♋ evaporates which softens swellings. Thirdly our fixed ☿ evaporates. He took 2 parts of this and ground it in the heat with 1 part of the coagulated precipitate mentioned before, until they formed an amalgam. He washed this with warm ▽ until the blackness went. He put the amalgam in a phial in the sand, covered it with ashes and applied the second degree of △, until it was red. Then he amalgamated it again with 3 1/2 parts ☿ rect., by grinding. This is very

146

easily done; the first time it is difficult, but if you grind constantly in the heat it is easy. When he had made this other amalgam, he proceeded as before.

Then he multiplied it again with 3 1/2 parts ☿ rect. This is done without grinding. He did this 7 times. He fermented it with the ninth part ☉ and was left with a liquid tincture unlike any other. For this tincture tinges 50 to 60 thousand parts. This is true. I have knowledge of the art from his own hand. If others had known this they would not have had to use ☉ or ☽ or other mineral materias to bind the ☿, because it can be bound as I have just told you. There are as many ways of purifying ☿ as there are days in the year, but this is the best. For my part, I prefer the ♁ and the sublimation. In this way the arsenic and sulphuric parts of the ☿ are driven out and burned, for the aqua fortis has a corrosive spirit which destroys everything. That which is impure cannot be made into the prima materia. You will find this tincture in every metal once you have purified it. This is such an old truth that it is the source of all the wisdom in the world. I should also say that the true dragon's blood is in ♂ and from this one of the best tinctures can be made.

Once Schulz said to me: Dear brother, do not neglect to work with ⊖ hois. Do not doubt that I

147

have revealed everything to you. One thing you must remember; the S.V. extracts the ⊖ red. When you have filtered the extract and have distilled off the S.V., pour it back over. Repeat this until you see a snow-white earth at the bottom of the glass. The extraction should take place in the warmth. This is the easiest way of doing it: Take red ⊕, put it in the cellar and let it liquefy to an ⚯. Pour the ⚯ into the S.V. It will dissolve more clearly and quickly. Otherwise do exactly what I have told you and you will succeed in our art.

But I said to Schulz: You told me about a particular of Gualdus's. What does this consist of? Schulz said: It consists ☿ viva, which is fixed with ♀ , ⊖ and ⊕. A solaric 🜍 is extracted from the noble ♈ which can be introduced into a fixed corpus. Then Gualdus took the clinkers and let them turn into ⚯ in the cellar. He added 2 parts common ▽ and precipitated it with distilled vinegar. Yellow 🜍 sank to the bottom. He took 4 Lots of this and 2 Lots of the ☿ made in the way I told you, and he sublimated it 8 or 9 times until the sublimate remained fixed with the 🜍 and melted into a red stone. If one part of this is added to 5 parts ☽ in flux, then it will be refined.

Shultz: Gualdus's augmentation is nothing other than this. The red precipitate must be imbibed with its own ☿ precipitate & rectified in the second and third Rotation. It must be imbibed with 2 parts and with 2 ½ parts, but not both together, but drop by drop over 5 or 6 times. The fermentation occurs with vinegar, burns it with 🜍 again after the vinegar has been extracted, extracts it with S.V. and he is left with a liquid ⊖ of ♂, which can tinge if a.a. ☉ is added. When the S.V. extracts red, the mass has to be burned with 🜍 again. He also took ☿⎯ made with ♋ and ☽, added the same amount of ♄ crude, and the same amount of ☽ and sublimated it another 3 times. He melted the sublimate to an ☊ which he left for 8 weeks in horse dung to putrefy. Then he boiled off the phlegm and he was left with an ☊ of ☿. He removed the phlegm from the water in the receiver and the ☊ of ☿ remained. It can unlock ☉ if 1 Lot of calx auri is added to 6 Lots of ☊ and left to stand for 8 days and nights in a warm place. Then he took 8 Lots of ☉ of ♂, 7 times he poured the ☊ of ☿ with the ☉ over it: 1 Lot every day in a warm place. When he had done this he put it in the sand, applied △ of the first degree, made sure the glass was well sealed and left it there for

149

14 days. It is refined by this 1 Lot added to 10
Lots.

Monte Schider said: I take crocus marti and
heat it until it glows. Then I rinse it with vinegar
or wood apple juice. I pour off the vinegar and heat
the crocus until it glows again, as if I were going
to smelt it. Then I rinse it again. I do this 12
times. I boil off the vinegar and pour S. V. R. over
the remains. I leave it to stand for 24 hours and
the S.V. extracts the ⊖ from the Ⓑ; (for the ♂
is not made into a ⊖ through the rinsing, but into
a Ⓑ, which is ⊖, and ♀, together; the ⊖ alone is
extracted through the S.V.). Then I boil off the
S.V. and am left with a ⊖ which can be used in the
same way as the one I described before. The ♀ of ♂
is also good, but only if it is separated from its
⊖ and if the ♄ does not accept it; for ⊖ dissolves
and ♀ is precipitated. To get the ♀ alone, take
some pressed grape juice made when the fruits are
unripe and only half grown. Let it stand in horse
dung for 14 days and then distill it dry. Then you
will have a spirit which is pure ⊖. This can be
used to extract the ♀ of ♂, because it must be
extracted using ⊖. This is another particular: Burn
♂ with ♀ common, extract it with the above
menstruum or spirit and it will extract yellow.

This is a sign that there is no ⊖ inside, otherwise it would extract red. Filter it and boil off the menstruum and pour it back over the ♂ as often as you see a yellow colour. Then take all of the extracted ♂⃰, grind it finely and pour the menstruum back over it. Boil it off 4 times. On the fourth occasion remove the phlegm from your spirit so that it is cleansed of all phlegm. Take 1 Lot of calx auri, pour 8 Lots of the menstruum over it, grind it for 24 hours, filter it and a corpus will remain in the filter but the a.a. goes through it. Pour this a.a. over the extracted ♂⃰ of ♂. Then boil off the menstruum and pour it over again 3 times. Then I add 1 Lot of this to 10 Lots of ☽ in flux and it is refined. The ⊖ of ♂ is made as follows: Take 1 pound of ♂ and dissolve it in 2 pounds of ♋ of ♀ which you can get from the smelter. Filter it, pour vg., (very gently?) over one measure of the solution (which must have been smoked until it was reduced by half) 3 measures of ▽ in a large sugar glass. Put pieces of wood in it and place it in the cellar. A ♑ will form which is bluer than the sky. In this ♑ is ⊖ and ♀. To separate them they must be calcined until they are red. Then pour filtered rainwater over and let it

151

boil. Filter it and coagulate it and you will find a white ⊖ and the ♀ will be in the filter. Add the ⊖ to a.a. ☉is and you will be able to refine the ☽. Unlock the ♀ some more, and you can try it with Luna. This unlocking can also be done with flours of Sal ammoniack, through sublimating until it is fixed and liquid at the same time.

Another Particular.

A certain alchemist took some iron filings which he extracted with distilled vinegar. He boiled off the vinegar and calcined it until it was red. Then he poured S.V.R. over it and then extracted it again. When the S.V. was white, he filtered it and boiled it off and he was left with a white ⊖ of ♂. Then he took 3 pounds of Hungarian ♁, and dissolved it in rain ▽. Then he filtered it and added 8 Lots of ⊖ of ♂ and a great deal of yellow ♀ sank to the bottom, which had become corrosive through the ♁. It should be left to stand in a warm place for 24 hours and then filtered, coagulated and the ♁ is no longer green as before. This is distilled per Retort. First a phlegm evaporates, then a sweet red ⚯. The latter evaporates with the second degree of △ (The receiver with the phlegma must be removed before the ♂♂ starts to rise.) He extracted the ⊖ ex ☺ with ▽ common. He took 8 Lots of this and 12 Lots of ♂♂, 2 Lots of Miller's ☉ and put it in a phial; leaving 3 parts empty. He put it in a Water Bath which was not warm enough for the ▽ to boil. He left it there for 8 weeks, then removed it and smeared the mixture on some silver lamellas and heated them until to glowing. He found ☉ instead of ☽. Monte Schider said: This doesn't happen because of the ♂, but because of the ♁ of ♀. I answered; this is true, but it cannot occur without the ♂.

Fauermann's Conversation

About the Lapis or Tincture.

Once I had a conversation with Fauermann and I asked him what the true matter consisted in, for there were so many methods of making the Lapis. He answered: There is one single thing from which the Lapis is made and this is in turn made by God Himself. If you take this thing from which it is made, and which is invisible, but unlocks visible things and brings them back to their first substance, that is; liquid and decomposed and then

153

fixes them again, then it is done. Then I said: Dear brother, this subject is so obscure; I cannot grasp what you mean. He said: Look, dear brother, the true matter, that which is fixed, the tingens, or the colour, is the ☉, and the other matter, the menstruum, is that which unlocks the ☉, this is how the ☉ is formed. But what is it formed from? It is formed from the 4 elements. What they are made of, God alone knows, but we do know what God has made from them. It is said that there are as many menstrua to unlock the ☉ as there are days in the year. This is true. But are they all equally good? No. Why not? Because a metallic menstruum is not as good as one not made from metal; which has been burned in the △ several times. For a metallic or mineral menstruum of the type we use, is our ▽, its primum ens is destroyed and burned in the △. It has to flow with ☿ until it reaches its primum ens again; if this is not precipitated through △, it is useless, because it is a raw, poisonous corpus as is also our ♄. I am not saying that such a menstruum is not good. It is good, but not as good as the true and most noble menstruum. No one takes any notice of this, but I am about to reveal it to you. Perhaps Herr Schulz has told you already, but I may be able to tell you a bit more than he has. So listen: The

☉ is made through the 4 elements. Which element is the noblest? It is the air which contains the other three and from which the ☉ is formed. For it hides in the mountain and the ☉ hangs onto the earth and forms a metal. But where can the menstruum be found? If I were to take the air, how could I catch it? I therefore do not take air, instead I take something in which the air has acted and which has been distilled and purified by the air. This is nothing other than a shower, or rain from a thunderstorm. Believe me, this is the very best materia in the world; I will tell you how you are to prepare it.

The Preparation of the Best and Finest Menstruii

Made From the Shower or
Rainwater from a Thunderstorm.

Take the shower water and let it decay and then evaporate it and reduce it by half. Put it in a wooden vessel until it stinks and is full of worms. Then pour it, together with the worms, into a flask and boil off the spiritum. Let the phlegm evaporate and pour the spirit over the remaining materia. Leave it to stand for 24 hours, well-sealed in a warm place. Then filter it. Take 8 Lots to 1 Lot of Miller's ☉. Put them in a phial and leave them to stand in a warm place until the spirit is yellow but the ⊙ sinks to the bottom and is black. Filter it and boil off the spirit and you will be left with a yellow ⊖. Now take 16 Lots of ☿ and pour 24 Lots of your spirit over it. Let it boil for 24 hours. Then take 8 Lots of your ☿ which is as clear as crystal, and 4 Lots of your ⊖. Grind them until the ☿ has swallowed everything and you will have an amalgam. Put it in ashes and apply heat of the first degree. In 5 weeks it will turn black. In another 5 weeks white and in a further 5 weeks it will turn the loveliest red. Ferment it with a ninth part ⊙ in a little flask and it will tinge 15,000 parts of all metals. If you want to ferment it, take only common ⊖ as I have taught you (i.e., the spirit is distilled and the phlegm has evaporated, the spirit boiled off, dissolved, filtered and distilled off again, then take the ⊖ which is left). Grind 2 parts of this ⊖ with 1 part of tincture, smelt it, and when it has changed through all the colours, ferment it again as before. Do this as often as you wish and each time you will be able to tinge 10,000 parts more. I recommend this to you.

Shultz said: Someone taught me this wonderful work: He calcinated ♂ very gently to ashes and boiled it with distilled rain ▽ and he was left with a ⊖. Then he dissolved ☿ in aqua fortis, precipitated it with common ⊖. The magisterium was left in the filter and the ⊖ went through. (This dissolves all metals and dissolves itself in every ▽.) He evaporated the water and sublimated the remains. In this way he was left with 6 Lots of sublimate from 1 pound of ☿. He took the 6 Lots and ground it with 5 Lots of the ⊖ of ♂ I taught you about before. He made it into an ⚇. Then he filtered it and found a red ⚇. He added 1 Lot of Miller's ☉ to 8 Lots of this ⚇. He sealed it well and put it in a common △, i.e. into a Lamp oven and applied the first degree of △ for 8 days, then the second degree until it turned black and so on, until it turned red. Then he found a red ⚇ which will tinge 3,000 parts ♄ and ☿ into gold. The ♄ must be smelted; the ☿ must be warm.

A Tincture from Chaos.

A ▽ is made as I have already taught. Take the clinkers and melt them with ♆ and a little ①. This makes another regulus. The clinkers are boiled out and the ♃ is extracted. All the ♃ from the regulus is extracted with ①, as long as it is coloured. Then the regulus is taken from the clinkers and is

157

finely ground with 1 part of this ♃. Then the kind of yellow ⊖ which was fixed on the first regulus, and was sublimated is added. Melt it and leave it like this for a good hour. (The crucible should be painted with chalk). Then throw in a little piece of crude ⊖, leave it to melt for several more minutes and then pour it out. You will find a red vitrum on the regulus. Keep it and the ⊖. Take 1 part of the regulus, 2 parts ♀. Grind them to an amalgam, distill off the ☿, apply a fierce △ until the regulus melts. Pound it, add 1 part of ♃, melt it and add the vitrum you find together with the ♃ic ⊖. Put the washed ♂ in the flux and melt it and add the pieces of ⊖. Then amalgamate it as before with ♀. But you do not need to use the ♀ which washes off. Do this 8 times. The final time the amalgam should be washed until nothing more comes off. Then it should be put in a phial. For fourteen days a gentle △ should be applied. Then the △ should be increased. Within 8 weeks it turns black. In another 8 weeks white crystals can be seen. If I want to speed up the process I increase the △ after the blackness has appeared. Then my work is as successful as Ruesenstein's.

Fornegg: I took a prepared chaos and poured it through 2 parts ♂. The regulus sank to the bottom. I added this regulus to the same amount of chaos again and amalgamated it with 2 parts ☿. I ground the amalgam finely and digested it for 14 hours. Then I distilled it off, smelted the regulus and added a little chaos to it. Then I amalgamated it with the ☿ from before. I did this 16 times and the ☿ was precipitated. Now the ⊖ ☉is had to be precipitated. This is how you do it: Take 1 part ☉ and 3 parts ☿ crude. Make such a tender amalgam from the two that you cannot feel it. Boil this amalgam for 6 hours in ⊡, then boil it off and grind it again. Finally, put it into a flask, pour some distilled vinegar over it and digest it again for 6 hours. Wash the amalgam until no more blackness comes off. Put this in a little flask, pour rectified spirit ⊡ over it and let it boil for another 6 hours. Then boil it off, wash it until no more blackness comes off and you will be left with ⊖ ☉is. Then I put the amalgam in a glass retort and I distilled off the ☿ (which is no use in our work). Then I took the ☿ which had been amalgamated with regulus of ♂ 16 or 17 times, and I amalgamated 1 part ⊖ ☉is with 2 parts ☿. I dried and ground

159

the amalgam finely. I put it in a phial in one degree of △ for 14 days. Finally I increased the △, to the third degree and in 6 months the blackness came, but only after great sorrow and danger before it turned white; for the phial must be smelted shut all the time, and there is always the danger that it will crack. But once the blackness has gone there is no danger any more. So when I get to the point where it is white, I can take 3 parts ☿ for imbibing. It is multiplied and imbibed and, when it has gone through all the colours, it is fermented white or red; whichever I wish. This applies only to ☿ and ♄, but it is a medicine which is just as good as Schulz's.

Monte Schider: I take my ♄, grind it to dust, smelt it and add 1 pound of rusty rails which have been heated to glowing point. I leave it melting for 4 hours and then I amalgamate it with 3 parts ☿, grind it finely, triturate it dry and then I put this ♂ in a wooden mortar. I grind away the ☿ and then put this ♂ in a phial. I apply a gentle △ to it in sand for 14 days and nights, then I increase the △ to the 2nd. degree. In 12 weeks it goes through all the colours. I ferment it with ☉, imbibe it 3 times with the ☿ separated during imbibing, and in 4 months I have 300 parts of this tincture. Before fermentation it can be used as a medicine to cure all illnesses.

Schulz: I take 1 Lot of ☉, dissolve it in 8 Lots of aqua regis, pour over 8 Lots of aqua fortis, and put it in a gentle heat for 2 or 3 hours. Thus the ♄ ☉ is precipitated. This is the a.a. ☉ is which many seek. It swims on the top like a little light. Fish it out, rinse it well and you have a good medicine. Boil off the aqua fortis and pour S.V. Tartarized over the ☺. Leave it in a warm place for 24 hours. Then you can filter and distill it and a ⊖ is left. Grind the precipitated ☉ with May dew ⊖ and you will have twice as much a.a. as you would if you were to take the fresh ☉. The ☿ must take on the qualities of a ⊖ through sublimation, otherwise it is no good. But if it is burned with ♄, it will also take on the qualities of a ⊖.

A common distiller took ☉ ore, ground it finely, poured stinking rainwater over it, covered it with a board and left it to stand in the sun until it was dry. Then he ground it up again. He did this 3 times. Then he poured enough ▽ over so that it covered it by a span. He put it in the sun and fresh air and left it until the ☉ was yellow. Then he sieved off the water with the skin and boiled off the ▽. He poured this ▽ back over the ore. He

161

continued like this until the ▽ turned yellow. Then he took all of his extracted ♂, added one part of it to 3 parts ☽ in flux and, it was refined. It can also refine ♀ and ♄ and other metals, but he only used it with ☽.

Ruesenstein's Chaos.

I make a chaos using the same amount of ♂ and ☌, without ⊖ or ⊕. I destine this chaos with arsenic. But the arsenic must be fixed. This is done as follows: I took 1 Lot of ⊕, 16 Lots of ⊖, 6 Lots of arsenic and melted it for 1/4 of an hour. Then I mixed this flux with the same amount of ⊖ and kept adding spoonfuls of this to the regulus. There must be 2 or 3 pounds of regulus to this mass. I continued to add the flux while it changed colour. I also added the ♁ from the clinkers to the regulus. I precipitated it with vinegar and added the ♁ mixed with an equal amount of ⊖. I covered it and left it melting for one hour and it was ready. The ☿ must be a.a. and its moisture must be taken away.

On our journey to Salzburg I once asked Schulz if a tincture could be made with a metallic substance without ☿. Schulz said: Dear brother, it is hard without ☿, although it could be done using

162

the dry method, that is without unlocking the menstruum. But you should know that the ☿ is the menstruum which opens the metal and then turns it into a primum ens again, that is into its first nature. Otherwise the metal or mineral will not be able to fix or tinge any other metal; for this the ☿ must be well a.a. I a.a. the ☿ with every metal, and with this ☿ I draw out the a.a. from metals. If I want to a.a. it with a 🜍 ☉is to a minera ☉is. I do the following: I make a regulus as I told you before. To 8 Lots of this I add 8 Lots of calx auri. I melt it without a flux. I amalgamate the regulus with 2 parts ☿ viva, grind it until it will go through leather, and it is properly a.a. mated, and this is the best a.a. of the ☿.

NOTE- The Columbae dianae is the regulus martialis.

Ruesenstein: In our work the ☿ is the man and the regulus is the woman. Fornegg: the ☉ is the man and the ☿ is the woman; sometimes we call both the ☉ and the ☿ the woman if we want to keep the work a secret. Ruesenstein: The △ is our regulus. Fornegg: the ☉ is our △. Ruesenstein: our △ is ☉, but it is a spiritual ☉, our 🜍s are far better than ☉, and if we also have a good flux we can

163

easily introduce our spiritual ☉ into ☽ and refine it. Shultz said: You must also remember that a corporeal ☉ must first be turned into earth, and only ☿ will make a tincture; for whatever is added to ☿, is only its earth and its fields from which the seed of ☿ grows. The ☿ must make your ☉ into fields and earth first. But our spiritual ☉ is earth from the beginning and can be made into ☉. Through ☿, with ☿ and from ☿ everything is made.

Ruesenstein's regulus made from Blackmahl; 3 parts of this and 1 part ☽ are melted together with a good part ⊕ Fixi, which is sulphuric, and 1/2 Lot of extracted ♃ from the Blackmahl. Also a few crumbs of ⊕ crude, and quite a lot of coal dust. The ☽ is refined through repeated smelting and the universal work can be made from it. Shultz once said: The putrefaction in your work is not completely black, only dark. But in my work it is as black as pine soot and within 4 weeks such a lovely growth can be seen from the top. In 14 days it collapses and it boils (that is in the putrefaction). In 4 weeks I already see the calcination. In 5 weeks the calcination has given way to the most beautiful crystals. The putrefaction is like a dung heap – afterwards little trees grow again which are also black, but a brown earth remains at the bottom. This is the calcination. The crystals are the interim philosoph. They represent the fixation. If I separate the crystals and put them in another phial without adding ☿ or a.a.a., and put them in the △, a red ♂ is formed which is my medicine of the first order. Ruesenstein: I took the crystals and amalgamated them with the seventh part of our ☿ and then I melted it again and within 4 weeks I had a red ♂ which I called my medicine of the first order. Schulz: That is good too, but it is not so fixed, and as a medicine it is not so effective. Now come the 7 multiplications: The first multiplication is when I add a.a.a., to the crystals. Secondly it is when I add 3 parts ☿ prepared to the crystals and imbibe it. Thirdly it is when I ferment the lapis. A universal work can also be made using ⊖ of ♂ and ♀ of ♄. I take a chaos ana., addition, grind it together with 1 pound of ⊖, 1 pound of ♃ and 6 Lots of arsenic, 4 Lots of ♀ and 8 Lots of ⊖. This is gradually poured into a crucible. The regulus should be melted eight times with this flux, the clinkers should be boiled out and the ♀ extracted and added to the regulus 7 times. Then I fix the ☉ with a common regulus, add it to the regulus with coal dust. Then the regulus must be amalgamated

165

with 2 parts ☿ common, distilled 7 times and then melted in the regulus as usual.

A Compendium Of

Various Wonderful Conversations

These 6 Adepts Had About Alchemy over a Period of 8 Days.

Amongst other things Schulz praises Theophrastus and calls him one of the heros of alchemy. He said that the beginning of his work was in ☿, in which he found something real; he described the work as follows:

The Work with Mercurio.

He took ☿ ⚖ ate, and three times as much wheat flour and separated it per Retort. It was very difficult to separate the ☿ from the fatty ⚭, and it caused him great sorrow. Then he thought up a different method. He dissolved it again in aqua fortis, distilled it off again and mixed the precipitate with 2 parts potash. He distilled it off per Retort and he found his ☿ shining and clear. He did this, as described, 4 times. The 4th time, when it was dry, he poured it into a phial and added a fifth part of Miller's ☉. He used a stopper to seal the top of the phial, put it in the sand and applied △ of the second degree. In 14 days the ☿ formed a little black skin. He continued until it was liquid, like pitch. Then he increased the △, but only by a tiny amount. In seven weeks he found a red ♂. He ground 2 parts ☿ purified, with it, as above, in the greatest heat. Then he smelted the amalgam again as described before and within 6 weeks the materia completed the putrefaction. Then he applied the third degree of △ and the ☿ melted and became a metal. He put it in aqua fortis but it refused to attack it, so he took one part, added it to ♄ in the cupel and he was left with the 7th. part of ☉. Then he took the remaining metal, amalgamated it with 3 parts ☿ purified, melted it as before, and it began to grow and turn black. This passed very quickly and it turned red. He took 1 Lot of this red ♂, added it to ♄, in the cupel, and he was left with a quarter of ☉. He imbibed it again with 2 parts ☿ purified, and it turned black and red very much faster. This remained fixed in the cupel. He mixed it with 3 parts ☿ again and continued as before, and was left with twice as much in the cupel, i.e., he got 2 Lots from 1 Lot. He did this another 4 times and finally could tinge 1000 parts with 1 part. Shultz continued: The arsenic poison is removed by precipitating and reviving the ☿. It can also be precipitated without any additions.

He also taught us about a medicine ex S.V., like the one we have already talked about, except that here you must make sure that the top part, or condenser, is always cold and it cooled with cold water; only 3 or 4 drops are taken from it, and it is a conservative for all illnesses. I make it by a better method: I take the strongest S.V. and put it in a tall phial. I put it in a pot full of water, sealed with straw or rags. I bury it for 18 weeks in horse dung and it turns as white as milk. (The feces are not separated.) I gently remove the phlegm. The ♉ which is left cures feverish illnesses, gout and other afflictions.

A Particular Collei ex Min. Solis & Menstruo.

He ground Mineral ☉ is finely and poured menstruum over it. A skin formed or the top which he kept removing carefully with a spoon. He dried this and ground ☿ ♎ with it which had been made according to Gualdus's method. It was precipitated white with ♁ and sublimated 5 times with the ☉.

Then it remained fixed and melted to a red stone. He added 1 part of this to 6 parts ☽ in flux and it was refined. The menstruum is stercus human., secretly taken from chambers and rotted with urine, and potash distilled from an earthen retort.

So you get a spirit, a volatile ⊖ and an ♋. The ♋ is separated through a funnel, the volatile ⊖ is rectified and the spirit condensed in the condenser, then poured over the mineral to a depth of a hand. In this way he got to know the nature of metals and minerals. This is how he came to make the tincture, for the remainder of the mineral is dead and none of it can be used. The smelters don't know this because they measure it by the △ which penetrates all corpora and introduces its noble ♃ of nature. If he wanted to make ☽ from ♀, he would take the mineram ☽ which has quite a lot of arsenic in it because arsenic is just volatile ☽ which will eventually turn into fixed ☽ and he would continue with the same menstruum in the same way as he did with the Minera ☉is. But he did not add the ☿ but added 1 Lot of its skin, which is thick and white like cream, to 12 Lots of ♀ in flux and its colour would charge to white, ☽ always remained in the cupel. If he wanted to make ♀, out of ♂ he did to the ♀ what he had done to the ☽ and ☉ and added 2 pounds to one hundredweight of ♂ in flux and it turned into ♀. In this way he nourished all metals. With these feces he introduced his 🜍 ☉is into ☿ and turned it into the finest ☉, when he added the dry skin to the ☿, 1 pound to 8 Lots. It didn't need a ferment, because it had a more subtle and open ferment in it. Dear brother, I answered, I like this method. It is true that there is a tinging 🜍 in every mineral which must not be attacked with any corrosive. I think, because this is done without corrosive, it must be good. You once told me something of Albertus Magnus which I do not understand exactly. I know about the ⊖ marino but did he not have something else too? Schulz said: He also worked with ☿. He purified it 8 times, then dissolved and precipitated it as described in the following:

Process using Mercury.

He dissolved 1 pound of ☿ viva in 1 1/2 pounds of aqua fortis and then boiled off the aqua fortis until it was dry. Then he mixed 2 parts potash and 1 part iron filings and boiled them off per Retort.

He dissolved what had steamed off again as before. He did this 9 times. Then he dissolved another 1 pound of ☿ in 1 pound of aqua fortis, precipitated it with common ⊖, washed it with fresh ▽ and sublimated it. He took 8 Lots of this and 4 Lots of the ☿ which he precipitated and revived before. He ground it until living ☿ could be seen. Then he put it in a little flask and left it to stand for 24 hours in a gentle warmth, then ground it with half its weight of fine ☿ and left it to stand again as before. He did this until he had ♀. When all was fixed and nothing more could be sublimated he took 8 Lots of it, ground it with 1 Lot of calx auri and applied the 2nd. degree of △, leaving it to stand until the flask was full of colours on top but black as ink at the bottom. Then he increased the △ to the 3rd. degree and within 4 weeks it was red and had the consistency of wax, but it was not like a stone or like glass, it was like a sponge. He ground this finely and added 1 Lot to 1 pound of warm ☿ and it turned within 1 1/2 hours to ☉. But to the Luna he added only 1/2 Lot and it became ☉. The ☽ hardly needs as much because it is fixed. Finally he augmented it as follows: He took 1 part of this of this ♂ and 2 Parts ♀ vivified. He ground them well and melted them until they were black and red. Then he took 10 Lots of this materia, ground it with 1 Lot of Miller's ☉ or calx auri, put it in a crucible and cemented it for 1 hour until it glowed brown. He tinged 30 or 40 parts. If he wanted to improve it he ground 3 parts ☿ with it as before. (Because the corpus of the ☉ is present it needs more ♀.) Then it tinged 50 or 60 parts, and so on. If you want to ferment and augment your tincture, you do not need to add as much to your materia prima as before, as at the beginning; only 2 parts. But if it is fermented and you want to augment it in quantity as well as quality then you have to add 3 parts to the prima materia. Then the prima materia unlocks the second or third materia completely, makes it spiritual and penetrating. Whether the prima materia is ☿, ☽ or ☉ it is worth the same: Nature is always happy with it. For if it comes from ☿ and it is treated so that it can tinge, the tincture is happier if it is added to other metals. If ☽ is made into a red tincture as Collero taught us, then the ☽, is happier if it is used to tinge things like itself than it would be if it were used for ♀ or another metal.

Schulz also uses viper's glass as a medicine to cure eating problems, and says it is black outside and white inside. He prepares it in the following way: He lets it wilt in the air, then he pounds it and extracts it with S.V. and then distills it to an extract. He does this 9 times and gives the patient a knife-tip-full every three days. Elleborus niger, not the common ore, although that is very good too, but the ore found on mountains and stony cliffs, has the power of all herbs and roots. It grows by itself but is planted by the May rain and can be found in the Autumn, green under the snow. Its leaves are almost as thick and sharp as a knife blade. Grind the roots when they are green, pour S.V. over it and then bury it in a well-sealed vessel for 14 days in horse dung. Then squeeze it out and that which is left is like cheese. Distill what is left, and the spirit will evaporate and the noble extract of the Ellebore will be left. Burn what is left in the cloth and extract the ⊖. When the ▽ is boiled to a liquor, put it in the cellar, add pieces of wood and the ⊖ forms crystals. This is added to the extract. It cures fevers and feverish illnesses. It should be added to wine or rose vinegar.

Collerus made a medicine using S.V. and rye bread. He added 1 measure of S.V.R. to 1 pound of bread and left it to stand for 12 weeks in horse dung. Then it was black in appearance. This was distilled per Retort and he was left with a phlegm, a spirit, an ⚯ and some ⊖ volatile. He poured

173

everything that was left in the receiver into a flask and boiled off the spirit and the ⊖ volatile. Then he put this aside and evaporated the phlegm, and he was left with the ⚇. He poured the phlegm over the ☉, extracted the ⊖ and kept it. The ⚇ was poured over the same amount of Venetian chalk, then the ⚇ was distilled per Retort and the extracted ⊖ was added. He gave the patient 2 or 3 drops and brought about wonderful cures. He gave the spirit to dying people, just 3 or 4 drops, and they recovered for long enough to make their last confessions. The ⚇ cures lung disease and frees people from all illnesses.

A Farmer's Universal using the Minera Lead.

He took as much ♄ as he needed, from a core, that is to say Marcasitam ♄, ground it finely and put it in a glass flask and poured distilled vinegar over it, he left it to stand in a warm place until the vinegar formed a yellow rivulet when the glass was moved. Then he filtered the vinegar and put it in another flask and then boiled it off gently, until it was quite dry. Then he put it in a glass refining retort, placed it in the open △, well-sealed, and distilled it. First he saw a clear spirit, then he removed the receiver, poured out the spirit and replaced the receiver and applied the 4th. degree of △ so that the retort was smelted. An ♋, like milk steamed off. Take the ☉ which is ashen grey with black spots and some molten ♄ at the bottom. Put the ♄ aside. Pour distilled rainwater over the remaining ☉ and leave it in a warm place for 2 to 4 days. Then filter the water and pour in your ♋ which immediately unites with the ☉ in the ▽. There will be 2 essences in the glass, the top half will be clear, like ▽, and the bottom red. The ♋ purifies and takes the ☉ of nature to itself. You can help this process: Pour everything together into a sugar glass and evaporate it until there is no more phlegm. You will be left with a liquor which is yellow. Pour the spirit you first distilled over it and a wonderful ☉ of ♃ will fall to the bottom. The moisture rises clearly and can be separated per filtrum. Your ♃ remains in the filter. Then take 2 parts of your ☿ ♎ and 1 part of this ♃, grind them together and put them in a flask in a gentle △ at first, then a stronger one. The ☿ is not sublimated, but remains fixed with the ♃ and you will see the red stone. This is the tinging stone, it can tinge all metals. Grind it finely, put it in

175

a flask, and pour 3 fingers of your S. V. R. over it. Put it in horse dung for 8 days and nights and then boil off the S.V. per retort. Put it aside, increase the △ and a red ♋ will evaporate. Grind the ☺ finely, pour over the S.V. which you boiled off and put it in horse dung for 8 days and nights, then proceed again as before. Do this 6 times. For the first three days a red ♋ will evaporate, the second three days it will be white. The farmer calls them the red and white flowers. They do not grow together, red grows with red and white with white. Do this as follows: Add 1 Lot of Miller's ☉ or fine ☉ plate and put it in a place so warm that you hand cannot stand the heat. The ☉ will remain whole to the 12th. hour, but it will be eaten in the 13th. and 14th. hours. Pour the white ♋ in a flask, add 1 Lot ☽ae lamellas to 10 Lots, and in about the 12th. hour the Luna will lie on the bottom like soot. If you now take it out and cupel it, it is fine ☉. If you leave it for 24 hours it will be opened up and will tinge white. Do this as follows: Melt the metal, add one drop to 1 pound which penetrates the metal in a second. This is the interpretation of the white and red lily or flower. He puts another lily aside, but says very little about it and does not let it see the light of day. This was his particular with which he started to work on the stone.

Another Particular of this Farmer.

He took 16 Lots of ☽, dissolved it in 1 pound of aqua fortis. He also took 1 pound ☿, dissolved in aqua fortis. He mixed the two solutions and precipitated the mixture with ⊖ which had been melted, dissolved, filtered and coagulated. He also added the same amount of rain water, he washed it with distilled ▽, then sublimated repeatedly until there was nothing left to sublimate. Each time he also ground it with the ☉ again until it became a stone. He added 1 Lot of this stone to 1 pound ♀ in flux and it turned into silver.

Philaletha's Particular made of Mercury and Sulphur.

He took 1 pound of common arsenic ground finely and mixed with 2 pounds of rye flour. He distilled it per Retort and first a spirit, then a thick, black ♋ evaporated. He separated this from the spirit, and to every pound of oil he added 1 pound

Of ☿ ♎ dissolved with aqua fortis and precipitated with ♁. He poured it into a flask, applied first a gentle △, then a fierce one and a spirit evaporated and nothing could be sublimated, it remained fixed. He saved the spirit which can be used to extract the a.a. ☉, ♀ and ♂. He removed the fixed stone which was as red as blood and which had melted to a red vitrum, ground it finely and augmented it 3 times, each time with the same amount of ☿ ♎. The third time the stone turns to liquid and is sublimated blood red in the condenser. He ground the sublimate with the vitrum on the bottom. He repeated this while it still sublimated, 7 or 8 times; and the stone is fixed again and a thousand times more noble than it was at the beginning. It is liquid like wax and enters the Luna like butter. It was refined, 1 part to 10 parts into ☉. Then the heat must be increased so that it melts because it has turned into ☉. The beautiful virgins mean the ☿ and regulus. His work using these is as follows: He took ☿ and ♀, made a chaos from them by mixing crumbs of ♁ with them. Then he poured it through 2 parts ♃ crude over the regulus which had sunk to the bottom, added some ♁ common and ☌; a mixture of equal parts so that it was cleaned of clinkers. Then he melted it in a crucible. Then he took an iron

mortar, poured warm water in it, ground 2 parts ☿ in
it and the regulus, and made an amalgam. He ground
this for 2 hours, washed it and distilled off the ☿.
He took the remaining regulus and ground it finely,
mixed it with twice as much ♀ crude, and poured it
through. He ground the regulus which he had just
poured through with ☿ again, as described, distilled
it off again after it had been ground for 2 hours.
He did this 9 times and the regulus turned yellow
and looked like ☉ and was liquid, like wax. This
can be seen melted in the retort before the ☿ has
evaporated off. After this the ☿ becomes tender and
leaves its skin behind which is very good. So take 1
part of your noble regulus for the 9th. time, add 1
part of the ☿ which you have melted often, make an
amalgam and grind it for 4 days and nights. Dry it,
fill a phial with 4 to 6 Lots of it and put it in
the sand, 3 knife blades above the cupel, as deep as
the substance is. The other bowl should be covered
with ashes and the second degree of △ applied. In 4
weeks it will be black, in 7 weeks it is completely
black. The △ must be constant until the materia
is red. Then grind 2 parts ☿ a.a. tum with it in 4
stages, and do as follows: Divide it into four. Take
1 part and grind it with the ♂ and put it in a

179

phial. Divide it again because 2 parts of it would not fit in a phial. Melt it again as before and the ♂ will be tender and liquid and will evaporate. When you see that there are many colours evaporating through the tube of the phial, pour the second part of ☿ in again. Then wait for this sign: When you see several flames shooting up then add the third part. The third sign is when the phial's tube turns black, then you should add the fourth part. Cover the bowl of the phial with ashes and continue with the △ as before. It will turn red twice as quickly as before. After the black phase it will immediately turn red. When it is red, proceed as before with 2 parts ☿ a.a. ti., like this: Divide it into 4, add it as before. The ♂ must not all leave the phial, only as much as the ☿ which was added to it. Weigh it for accuracy. Then continue as before with the △, repeat it 4 or 5 times. Three times is what it needs to gain the power to overcome the fixed metal. If you only do it three times, take the red King on the third occasion and add it to a Queen, i.e. 8 parts of a fermentum ☉is or ☽ and it will turn white with ☽, red with ☉. But let me tell you how it must be fermented: Put your tincture or red ♂ in a smelting crucible which has been painted with chalk. Then the tincture starts to rise, like a tree, quite red. Add an eighth part of your ferment and immediately it will sink to the bottom. Then leave it to cool down and grind it finely in a glass mortar and it will be the colour of sulphur. Fill a little ivory box with it and it will tinge 1000 parts if it has gone through 3 rotations. If it goes through 5 or 6 rotations it will tinge 10 or 12 thousand parts every time it goes through a new rotation. After the sixth rotation it tinges 10,000 parts more.

I answered: Dear brother, if a man could be sure that this would happen he would achieve his aim and reach the end of his work. Schulz said: No, because you must first know what ☿ a.a.tus is. You also need to know what the Columba Diana which animates the ☿ is, and finally you must know the poisonous dragon from which the noble river of pearls springs. I answered: Dear brother, this is why many alchemists work until their death but achieve nothing, ever though they read the best books. The processes appear to be described clearly, and one would think it impossible to fail in carrying them out, yet one does fail, and knows not why.

Schulz said: Dear brother Ruesenstein, if everything were clearly explained, our art would not be rare but common. But it remains a secret, and we can only succeed by studying and practicing. But in order that you might know that I am telling you everything, and not withholding anything from you, I will unravel the first knot for you, a knot so dark that no one could understand it who did not practice the art. What is the poisonous dragon from which the river of pearls comes? It consists in the following: The ☿ is prepared so that the regulus can be a.a. with it as to the second. But the ☿ is a.a. as follows: Take 1 pound of aqua fortis and the same amount of ☿ viva, which is dissolved in it. Boil off the aqua fortis and grind the precipitate left behind finely. Do the same to the same amount of ♀ common, and ☉. Put it in a glass flask, apply a sublimating △, and the ☿ is sublimated. This is called the poisonous dragon and in this nature's noble river of pearls is hidden. Add 2 pounds potash to 1 pound of this sublimate and distill it per retort. Then you will have 24 Lots of ☿ from 1 pound. This is the noble liquid of nature. He does this first wherever he makes it into an a.a. to add to the Columba Diana. It is done in two ways: The first is the clear and bright regulus, the other is ♓. The first is added to the regulus when the ☿ is prepared for the work. Then the ☿atum is added so that it is a. But the other Columba ♓ is made as follows; dissolve and precipitate it in aqua fortis

182

when it is washed and dry. Then amalgamate it with four times the weight of river of pearls in a glass or stone mortar. Grind this amalgam for 4 days and nights with distilled rain ▽. Pour off the ▽ and distill it per Retort, grind the ☿ again with the ☉ as before. Then distill off the ☿ again. Do this 4 or 5 times until the Columba Diana is dead, and is brown instead of white, it is a dead earth, it evaporates from the cupel. But the ☿ is well a.a. ted. If you pour it off, it leaves behind a skin. Distill this skin and it will evaporate. Use this as I told you before and you will discover the art. But without this point your work will be in vain. But dear brother, I said, I thought the poisonous dragon was ☿! Yes, said Schulz, firstly he calls ☿ the poisonous dragon and ♂ the yellow lion. They both tear at each other, they both die at the same time.

But the other dragon, the double ☿ touches the 2 dead ores and bears a beautiful virgin which is the tender and lovely regulus which he uses as has already been described.

A Particular Philalethae.

Schulz said: The particular of Philaletha consists in ☿ reg. and Mineram ☉. He fermented the

183

☉, took 1 pound of ☿ and 18 Lots of minera ☉, which was tested in the cupel. He powdered it and put it all in a smelting crucible and melted it. Then it formed a skin which had to be fished out with an iron spoon because it is the slimes' quartz, which can form glass. When the minera had melted, he prepared one pound of the best ♂ in the △ until it glowed and then he added the melted mineral. He covered the flask tightly and left it melting for a good hour before pouring it out. He took the one pound of the regulus and 2 pounds of ♄ crude, ground them and melted them together. When he poured this out he had a yellow regulus which he kept. Then he took 2 pounds of clinkers and mixed them with 2 pounds of crude ♀, and 1 pound of ⊕ and 1/2 pound of ⊖, ground them all together and melted them in a crucible. The liquid ♃ sank in the form of a fiery regulus to the bottom. He also kept this and then boiled out the clinkers to get rid of the bad earth from the ♂ and the ♄, and also the remaining ♃ which the flux had attracted. When this was done he filtered the ▽ and coagulated it to a ⊖. When you have 2 pounds of this, grind 24 Lots of common ♃ with it and do the following: Grind the regulus which fell from the clinkers, gradually add it to a crucible with the flux in it and melt it. When all has melted you will find a true stone just like glass and almost no regulus, or perhaps just a little, on the bottom. Grind the clinkers finely and boil them out in an iron pan. Filter them while they are boiling, otherwise the ♀ will solidify like liver and not go through the filter. When you have filtered it, the ♀ solidifies like a liver which must be removed and washed with ▽. Pour distilled vinegar into the remaining ▽, and the remaining ♀ will be precipitated. Then filter the ▽ and the ♀ will remain in the filter. This must be done to the others too. Take 3 of 4 Lots of the first King that you poured through when you have cleaned off the clinkers. Add 1 Lot of the ♀, finely ground, to 4 Lots. Place them in a crucible, painted with chalk. Put it in a smelting △. At first the heat should be gentle, so that the ♀ has the power to open up the regulus. Apply this gentle heat for one hour, then smelt it and pour it out. You will find a very small stoney regulus, full of ♀. Grind this King again and mix it with the ♀ from before. You will find a red vitrum or the King which must also be ground with it. Proceed as before. Keep repeating it until you have no ♀ left and it is ready and fitting for a particular work. Then he does the following:

He takes ☿ ♎ as I taught you, prepared with ⊖
from aqua fortis. He revives it with 2 parts potash.
He takes 8 Lots of this revived ♀, and grinds 1 Lot
of his calx auri with it. He grinds the amalgam for
24 hours and then distills the ♀ from it. The ☉
☉which remains in the retort is compact and must be
dissolved with aqua regis, which must then be
distilled off and you are left with calx auri. He
added his fine ☿ to this and it absorbed the calx
auri again. He ground it again for 24 hours with
great force (in an iron mortar with warm vinegar)
and then he distilled the ☿ from it, but not
completely. The ☉ remained a bit white and could
easily be amalgamated. He amalgamated it with the ♀
which had been distilled off and ground it again as
before. He did this 7 times and finally a black ♂
could be found in the retort instead of the ☉. This
is very light and it is useless, for the heavy
♁ ☉is has become ☿ through the ♀. This can be
used. He proceeds as follows: He takes 8 Lots of the
regulus he made before and adds to it 16 Lots of the
☿ prepared in this way. He makes an amalgam which he
grinds finely a puts in a phial, as long as he does
not have more than one pound of it. He puts it in
the sand, one finger's distance above the cupel, the

186

whole bowl covered with sand. He applies a gentle △
for 4 days and nights before increasing it to the
2nd. degree until the end. Within 14 days the
materia turns black, and in four weeks the blackness
has gone. For the noble ♄ ☉ is which is in the
regulus attacks the ☿ with force, which has already
been dulled through the ☉ and it can easily fix
itself. After the blackness it turns red within 5
weeks. This can be used.

Philaletha first added it to ☿ viva; 1 Lot to 6
Lots. It melted to a vitrum and the ☿ went away. He
did not know why it had gone wrong. He took 1 Lot of
it, fermented it with a tenth part pounded ☉, then
added 1 quarter to 6 quarters of warm ☿. It was like
butter. Then he broke open the crucible and he found
very brittle ☉ inside, which could be ground to a
♂. He thought that he must have taken too much of
the tincture, which is surely the case. Then he took
1 quarter of this and added to 10 quarters ☿ which
was warm, and only then did he find true and noble
☉. Then he fermented everything and refined the ☽
in the same way as the ☿. He also augmented it as
any other tincture. But it couldn't be used as a
medicine because the ☿ made it poisonous, as did the
arsenical ♄ which was in the regulus. But this can

be used in this work, because it can overcome the ☿.
Then he took the fermented ♂, and ground it with ☿
revivified, from the sublimate without the ferment
☉is. Then he melted it again as described before.
It had gone through the stage of putrefaction in the
same time as before. He did this as often as he
wanted. Each time he found that it would tinge 10
parts more. For the ☿ is a key which cannot only
open up metals which are open already, but it can
also open up all fixed and compact metals and fix
them. Although it also appears as a white ☿, it
contains a tincture which improves the ☉. Because
if you precipitate it with ☽, or other metals, a
tincture is found which is as red as blood. But
why can it not be bound, why does it have a ♄
inside, and why can it be bound with the help of a
☉en or ☽ a.a., but not without it? The reason is
that it contains an arsenic poison. If it is not
robbed of this, then you can add what you wish, but
you will never bind your tincture. But if you
separate the poison from it, you need not add
any ☉ or ☽, it can fix itself, and needs only a
corporeal ☉ or ☽ as a ferment. This fixes the
tinctural ☿. If the fixed metal is introduced, you
have done everything. The answer: I have learned

that ☿ contains a wonderful tincture and I have often been surprised that it is so difficult to bind.

Shulz: Dear brother, just as ☿ is a bird which makes fixed metals into ☿ and won't let them be separated from him unless it is precipitated. In the same way the poison is attached to him in that it carries the fixed ☿ and other fixed metals with it into the smoke. Dear brother, you should know that Philaletha also had another particular, of Minera ☉is. Really! I said, is it made with corrosives or through distillation or through fulmination? Yes, said Schulz, through fulmination, as I will now explain.

Another Particular.

He takes one pound of a good solaric mineral which only carries one red ♁ inside it. He grinds 2 pounds of common ♁, 3 pounds of crude ♄, 1 pound of ①, 1 pound of common ⊖ with it, and heats a crucible until it glows. Then drop by drop he adds the mixture, covers the crucible with a lid and some coal. He leaves it to melt for an hour before pouring it out and grinding it finely. Then he pours boiling water into a pan and leaves it boiling for a good hour. Then he filters it while it is still boiling, then filters the filtered ▽ again. When it is cold the fixed common ♇ will remain in the filter, but the ♇ ☉ will be in the ▽. He precipitates this with vinegar, filters it and the ♇ is left in the filter. He washes these two ♇, dries them very gently, for they are very liquid. With one Lot of this and 4 Lots of ☽ makes s.s.s., cements it for 4 hours. Then he melts it together and in the core are four Lots. This was Philaletha's best work from the beginning, just Like Monte Schider's, for he destroyed the ♂ the same way. But I would like to tell you, dear brother, that in Monte Schider's work, the main point has been withheld up until now. For I have not told you everything, and he himself usually leaves something out of his tracts, as other authors who write about him also do. But this fulmination just reminded me that I had better warn you and tell you about it. He took Mineram ☿, and removed the coarse earth. For when this minera is put into the △ and is made into a fixed Corpus, the noble ♇ which it needs disappears. It goes through all the colours as described, but no one will be skillful enough to perform Monte Schider's work if he does not do the following: He took 1 pound of this Minera and added 1 pound of common ♇, 2 pounds of crude ♀, 1 pound of ⵔ and 16 Lots of common ⊖

190

to strengthen the flux so that it can attack the Mineram. He ground it all with the Minera and gradually added it to a glowing crucible. He covered it and left it melting for an hour and a half. When he poured it out he found a little fixed ♀, which was really wild and could not be used. But he pounded and boiled the clinkers, filtered it while it was boiling, and filtered the filtrate again. Or he removed the red liver of the ♁ but precipitated the other ♁ in the ▽ with vinegar, filtered it, washed it and dried it. Then he did the following with this dried ♁: He took the Mineram ♄ which is attracted to ☽ ♁is and can be recognized if the Minera has a white crown. He took 16 Lots of this and 16 Lots of the other ♁ too. He ground them, cemented them in a crucible for an hour, then smelted them with the fiercest △ until the molten ♄ covers the Minera ♂ like water. Then he poured it out and found a ♄en King which has yellow threads around it and some brown vitrum over it which is useless. He proceeded with this King as is explained below.

Monte Schider's

Particular ex Vitriol Albo.

He takes white ♁ which can be found in ☉ mines and which is called ♁ ☉is. (It contains red and black ♃). He grinds it finely, distills it per Retort with strong △ for 24 hours and he is left with a spirit. He takes 1 pound of the ☉, which is as red as blood and grinds it with 1 pound of common ♃, 2 pounds crude ♀, 1 pound ☽, and 16 Lots common ⊖. He adds it to a glowing crucible and leaves it to melt for a good hour. Then it is poured out and boiled out as explained below and he is left with a quantity of red ♃. He takes 4 Lots of this, 8 Lots of the spirit from before and he extracts the ♃ ☉is. Then he filters the spirit through a cotton cloth so that the rest of the dead earth stays in the cloth. He puts ☽ lamellas in this ♁, so that the ♁ just covers them. Then he puts it in a gentle warmth for 24 hours and the lamellas turn as black as coal and spongy. He dries them and heats them until they glow and they look like the colour of ♃. Then he smelts and cupels them and he has the finest ☉. There is a quicker way of doing it: He takes as much of the white ♁ as he wishes, grinds it finely and takes a glazed pot, fills it with 4 fingers of potash and pours over ground ♁. Then he covers this again with potash and sends it to a potter or tiler to fire it in the kiln. He finds a red lump. He takes out the ♀ which has fixed inside, grinds it to a powder and fulminates as described in the above work until he gets the ♀. He takes one part of the rinsed ♀, 2 parts calx of ☽ precipitated with ♀, mixes them and puts them in a smelting crucible which is sealed well. It is cemented for 3 hours, then smelted and poured out and inside he finds 3 Lots of ☉ in the core.

Another Particular ex Eodem.

He took this ♁, distilled it as described above, with the strongest flame of △. He ground the ☉, boiled it, filtered and coagulated it, and he found a noble Arcanum duplicatum. But it wasn't a real Arcanum duplicatum, because it was only made of ♁ and not ☽, but is many thousand times more noble and is more use in medicine than the Arcanum duplicatum ex ☉ aqua fortis. This is used for all fevers and can also be given to those suffering from feverish illnesses, for it causes a heavy sweat and rids the body of gall and mucus. But enough of all

193

this. He dried the ☉ which was well boiled but had no ⊖, and put it in a retort. He poured the evaporated ℞ spirit over it and then boiled it off again. (This must be done in a glass retort).

Then you must apply a fierce 🜂[15]. When the spirit has evaporated, let the retort cool down, break it open, and you will find the 🜍 of ℞, much redder than before. He grinds this red ☉ to a subtle ♂. He puts it in a fresh retort, pours the evaporated spirit over it again and then boils it off again. He does this 3 times and the 🜍 is open, liquid and fixed. He takes 1 part of this 🜍 and grinds it with 3 parts calx of ☽, precipitated with ♀ plate, and cements the mixture in a smelting crucible for 5 hours. When it has all melted, he finds 2 1/2 Lots ☉ in the core. Monte Schider also found this. Schulz continued: Dear brother, before I tell you another of Monte Schider's I must quickly tell you about an rare particular of Colleo, which doesn't take long. The spirit of ☿ which Monte Schider used in his work reminded me of this.

---

[15] Here, Hans Nintzel used the symbol 🜍, sulphur, in error. -pnw

A Brief Particular.

Colleus takes one pound of arsenic which must be whiter than snow. He grinds it to make a subtle ♂ and dissolves it in 2 parts aqua fortis, made of 2 parts ♄ and 1 part ☉[16]; anything else will not dissolve it. He takes 1 pound of ☿viva, dissolves it with 2 parts, takes 16 Lots of ☽, and dissolves it in the same aqua fortis. He mixes these three solutions, boils off the aqua fortis per Retort, grinds it to a ♂, puts it in a flask in the sand and adds the following ♋ to it drop by drop; Take crude ♀ ----[17] & hic erat spatium vacum[18], & omnissus processus olei, deinde sic pergit; the last three times you should apply a sublimating △ and a white, fixed stone will remain at the bottom which is liquid, like wax. This happens after 15 boilings and rinsings. Then it is ready, he adds 1 part of this stone to 8 parts of ♀ in flux and he has ☽. He also fixed the ☿ like this: first he dissolves 1 pound ☿ in aqua fortis and extracts it per Retort. He adds 8 Lots of this precipitate to 1 Lot of the stone, grinds them, cements them in a sealed

---

[16] The original text says Gold, but this is wrong. –Hans Nintzel
[17] The original text is illegible here. Hans Nintzel
[18] Literally, "Here was an empty space." -pnw

crucible for 2 hours in a circulating △. When they are smelted he finds fine ☽. But he only gets 6 Lots from 8 Lots, because 2 Lots disappear. Dear brother, now I will tell you about a spirit of ☿ per distillation which Monte Schider made.

A Particular cum Spiritu Mercurii.

He takes ☿ sublimated with ♄ and ⊖. He puts them on a grinding stone and puts it in a cellar where he lets it melt to a ▽. He filters this, takes white ☿ ♎, as has already been explained and grinds it. He puts 8 Lots of it in a retort and pours 16 Lots of the ☿ ♎ ▽ which has melted in the cellar. He fits it with a large receiver and distills it, first gently, then with a stronger sublimating △. A volatile, strong spirit evaporates and is sublimated in the neck. The sublimate and the spirit are returned to the retort. When this has been done 6 times, the whole sublimate will have become a spirit. He takes 8 Lots of this and one Lot of ☉, beaten thin. He lets it gently digest for 24 hours and the ☉ dissolves in the spirit of ☿ and the spirit becomes golden-yellow. He puts as many ☽ lamellas in this golden-yellow essence as the spirit will cover. The remaining spirit which does not penetrate the ☽ must be distilled off per retort. If it leaves the ☉is Corpus with the Luna, you have to ferment the Corpus ☉is again with the sublimate of ☿. Otherwise it will be useless. The spirit with the ☽ must also be put in a warm place for 12 hours. Then it is distilled from the ☽ with first a gentle and then a strong △ until the retort glows. When it has all evaporated, let it cool down, break open the retort and you will find ☉ instead of the ☽. Theophrastus did something similiar, but he didn't call it a spirit, he called it an ♋ of ☿. Here it is:

Particulare ex Oleum feu Spiritu Mercurii.

He dissolved ☿ in aqua fortis and precipitated it with ♁, sublimated it and he had some white crystals. He poured 16 Lots of aqua regis over 8 Lots of the crystals and distilled it off. Then he applied a sublimating △, poured it over again and continued to do this until the sublimate had completely evaporated with the aqua regis. This he called ♋ of ☿. This ♋ he put with ♋ of ☽ and ☉. He fermented the ♋ of ☽ or ♋ of ☉ and added it

197

to the ♋ of ♀. If he wanted the ♋ of ♀ to be a red tincture, he made it into an ♋, with the ☉. If he wanted to make a white tincture he used ☽. He took 16 Lots of the ♀ ♋, added 4 Lots of thinly laminated ☉. This was quickly dissolved by the ♋ of ☿. He distilled the spirit off 8 times and the ☉ evaporated per Retort and left a black dust behind. This is called the ♋ ☉is. It is fermented again, and then 8 Lots of it are added to 1 Lot of fine calx auri or ☉ lamellas and digested gently. The ☉ dissolves quickly. If he wants to tinge, he takes as much ♀ as he wishes, dissolves it in aqua fortis and boils it off to a liquor, pours it into a flask and for every 8 Lots of ♀ adds 8 drops of ♋ ☉is to the liquor of ☿. The ☿ sinks to the bottom in the form of a black ♂. Then he extracts the liquor per Retort gently, applies a strong △ and the corrosive part of the aqua fortis evaporates, which performs a wonderful separation if ☽ lamellas are added. For it also takes something of the ♋ ☉is with itself. A yellow Corpus remains in the retort. He cupels this and finds ☉ instead of ☿. This is what he does with the ☽: Take 16 Lots of your ♀ial ♋, add 4 Lots of ☽. Put it in a warm place and the lamellas will melt and turn black. They take 24 hours to dissolve. If you take them out after 12 hours and cupel them then you will have 1 Lot of ☉ instead of your 4 Lots. The other Lot goes up in smoke. But if you wait 24 hours they will dissolve. When they have dissolved, pour the substance into a retort and distill off the ♉ again. Proceed with the ☉ as before and you will have the white tincture. It must not be added to the ☿ alone, but to the ☽; the ♉ ☉is to ☽, but the ♉ of ☽ to copper or ☿, and nothing else. But the ♀ must not be dissolved in aqua fortis. Instead it must be filed. Then add 8 drops of ♉ of ☽ on it and it will turn into ☽. You would fail here if you were to proceed like this. Do the following: Take 8 drops of your tincture in three spoonfuls of ▽. The ▽ will immediately change colour and swallow the tincture. Pour this ▽ over 8 Lots in a sugar glass, put it in a warm place so that the phlegm evaporates. When it is dry, cupel it and you will find ☽ instead of ♀. When the ♉ of ☿ is ready you can take 16 Lots of ♂ filings and then proceed as with the ☉ or ☽. But it has to be fermented with ☉ otherwise it will not be able to tinge. It can do this with ♀, but both must be fermented with ☉. With ☽ the same applies, but you don't have to ferment it. It will not become ☉ or ☽. It will be a hermaphrodite, i.e., mixed up, for if it is brought to a tincture they don't tinge white, but red instead. If the tincture is red, it is wholly red, but if it is white, it is half red, half white.

Then I said to Schulz: Dear brother, it is time I left you and returned home to my people. Please advise me, dear brother, what things I should always have on hand. Schulz said: First I recommend ⊖ hois and the mineral ⊖, which is the Basilius's and Monte Schider's ℞, for you must learn the nature of our art, which consists in a single thing; in the principle of all things, that is ⊖. For this is why we need the mineral ☿ so much; it is a pure ⊖ ☉is. It can be seen clearly enough when it is sublimated that it is nothing other than a pure ⊖. This is why we need it for everything in the metallic art. Monte Schider makes it and sees it as lord above all things when it is a ⊖. I answered: Dear brother Schulz, I am with Monte Schider; for I too praise ☿ above everything else, for I have found many wonderful things in it, e.g. my Augmentum ☿ii, ☉is and ☽ae. I purify it, but I do not make it into a ⊖, but I unlock it, so that it is freed of its

arsenic poison. Schulz answered: You are right, dear
brother, remain steadfast, you will learn many good
things; if you just take its spirit it will
accommodate itself to many things. I continued;

☿ is a robber and a giver. When it is purified it
gives, but when it is impure, it robs volatile and
fixed metals which are added to it. I know it only
too well. There can't be many people who have driven
more ☿ into smoke than me! Dear brother, said
Schulz, driving ☿ into smoke is no art, but fixing
is very difficult. There must be many who drive it
into smoke. I recommend you, if you want to make a
universal, to prepare the ☿ first with aqua fortis
or ♄, these are the two best preparations and
purifications for ☿ and in this is the whole art of
all metals; for if it burns through the ♄ and is
revived, it loses its arsenic poison and takes on
a mineral ♄ which it reeds, for without it it cannot
be fixed. I answered; dear brother, I have taught
how far ☿ can be ground and washed with another
fixed metal, so that it is purified and animated?
Schulz; Yes, it is a., it takes on some volatile
metal which it draws out with its arsenic poison and
robs the fixed metal of it, but it is a., not
purified. In order to purify it and rob it of its
poison, it must be ground for a long time or the

fixed metal must be unlocked. It must be added to it
in a ferment, for the metal must be joined to liquid
☿ and made into an amalgam. When it is put into the
△ it is powerful enough to unlock it ex fundamento.
I recommend this to you above all particular and
universal works of metals. This is the most
important point for you to watch, without this there
is none other which will fix it without a tincture.
I also recommend you the ⊖, which, in so far as
they are purified, putrefied and made liquid through
dissolving and coagulation, contain wonderful
dignities. You can do a lot with them, especially
with ⊖ hois; for when it is made liquid, give it a
Fermentum ☉is or ☽ae and it will give off a
wonderful tincture and all you have to do is wait.

I said; dear brother Schulz, all this would be
very good, if only what was written and said were
true. If it were one would soon be richer than the
Roman and Turkish Emperors together! Schulz said; I
cannot guarantee that everything is true, but I have
not withheld anything from you. However, authors
sometimes hide a part of their work here and there,
and the alchemist must find it through hard work and
experiment. So not everything that is written is
true. But so that you do not go wrong, I advise you
to work only with ☿ in your particulars and
universals. You can also prepare a ⊖ hois. I assure

you that if you purify the ☿ as described, you will
fix in a particular and a universal. But you may
rely on the short particulars of Gualdus,
Theophrastus Paracelsus, which I have taken from
their own hand written works, and I have not left
out a single word.

As I was leaving my brothers in Salzburg to go
home, they gave me a number of different teachings
and warnings which refer to laboratory work and our
art. Again they recommended ☿ before all else, and
said that it was God's will that I use this first,
and purify it first to the best of my ability with
aqua fortis or ♃. But Koller said; There is no
better purification of ☿ in the world than that with
aqua fortis, dissolved, precipitated with common ⊖
and sublimated, revived, dissolved again, sublimated
and revived, and that is the best purification which
can be found between heaven and earth. The reason
for this is that the aqua fortis dissolves it, and
through the dissolving and precipitating, the ☿
becomes more volatile than usual; in fact nothing
could be more volatile. Because of this, and because
of the sublimating, it loses its poison which
remains in the ☉ of the ⊖, and nothing is
evaporated except the good and noble ☿ which is the
pure essence in all metals. The ☿ is strengthened

through it, but its poison is weakened, burned and destroyed.

Just at this moment, as I was about to get on my horse, Schulz said: Dear brother, let me tell you one more thing, and it is the best thing, it is how you can make a quick tincture:

A Quick Tincture cum Menstruo
ex Water pluv. Tonitr. & Grandine.

Take rainwater from a thunderstorm or hailstorm. Evaporate the ∇ until it has reduced by half. Then put it in a wine barrel under a roof where the air can reach it for 4 or 5 weeks until it stinks and is rotten. Then pour it into a brandy kettle and distill it as you would a brandy. When you cannot see any more spirit rising, then pour more of your ∇ into the kettle. When you have distilled everything, save your spirit, filter the remaining ∇ and evaporate it in a sugar glass to get out all the phlegm, until it is dry. Then you will find a rubber-like essence. Pour as much of your spirit over this as necessary and leave it to stand in a warm place for 2 or 3 hours, well-sealed, and the spirit will extract the $\ominus$. Then filter it and extract the spirit and you will be left with a pure $\ominus$. Do the following with it; Add 1 Lot of

Miller's ☉ to 2 Lots of your ⊖ in a glass mortar
and grind until the ☉ charges colour. Then moisten
it with a sprinkling of the extracted spirit and
grind it until the materia is black and bubbles up
like foam. Then pour more spirit over it so that the
⊖ can be dissolved with the a.a. When it is
dissolved, filter it through a filter paper, pour it
in a flask and extract the spirit. The ⊖ remains
behind with the a.a. Then take 1 pound of your ♀,
put it in a glass mortar and pour one measure of the
spirit over it. Wash it until it looks like a mirror
and is as blue as a plum. As soon as the ▽ turns
black, pour off the ▽ and evaporate it again into
the condenser. Then pour it over it again, and wash
it until it is black again. Do this 3 or 4 times
until the ☿ looks as I told you. Then dry the ☿.
Take 8 Lots of the ⊖ I have just taught you about
for every 16 Lots of ☿. Put them both in a glass
mortar and grind until you have a fine amalgam.
Put half of this in a glass phial, place it in a
philosophic Lamp oven and proceed as I have often
told you. Then you will have a tincture for 20
parts. Augment it with an amalgam; each time one
part tincture after fermentation and 2 parts
amalgam, and it will tinge many more thousand parts.
Then he gave me a kiss, wished me much luck on my

journey, and promised to write to me often, especially if he learned anything new. The others did the same, and I kissed them all and they begged God to look after me. And so in the name of God I mounted my horse and set off home, which I reached happily and peacefully in 5 days.

A Particular ex Antimony & Lumbricis
una Gutta ad 1 Pound Mercury vel Lead, which Ruesenstein was Sent in a Letter.

He wrote to me and said I should take 8 pounds of the mineral ♁, grind it finely and then take a quarter of a jug of rain worms, leave it to stand in a retort in horse dung or in Balneo for 4 weeks and then I should seal the retort, and put it in an open △, together with a large receiver, sealed with luto. At first I should apply a gentle △ until drops began to fall, then I should increase it, each time by a degree, until the third degree, and I would be left with a spirit, an oil and a volatile ⊖. I should pour the contents of the receiver into a flask and rectify it. I should save the spirit and the volatile ⊖. I should remove the ☉ of the worms and pour distilled rainwater over it. Then I should boil it out, filter it and coagulate it, and I would find a white ⊖. This salt I should add to the rectified spirit. Then I should do the following: I should put my fine ♂ of ☿ in a glass flask, and then pour this spirit over it, then leave it, sealed properly in a retort with a large receiver, in horse dung for 8 weeks. First I should apply a gentle △ and a clear spirit would evaporate. Finally a blood-red ♑ which he called acetum of ☿ would evaporate. Everything should be rectified in a flask, and the ♑ or acetum of ☿ would remain behind. I should put 8 or 9 Lots of this ♑ in a phial with 1 Lot Miller's ☉, smelt it shut on top, and place it in a secret Lamp oven. It should stay there for 8 weeks and in a short time colours will be seen. Then it would turn black as ink, then red. But I should continue with the 2nd. degree of △, then remove the phial and I would find a red ♑. If one drop is added to 1 pound of ♄ or to warm ☿, they would be tinged into ☉.

Arcanum ad Gold Putab, ex Urine & Gold.

He drank nothing but wine and then distilled his urine as soon as he had collected it, to extract the common ⊖. He put it in a barrel, covered it and

207

put it under a roof to rot for 12 weeks. Then he
extracted the spirit and smoked off the ☉ in a
sugar glass and the true ⊖ remained behind, which
was not in cooking salt, but can be found in wine.
He dissolved this ☉ with rainwater, filtered and
coagulated it gently and he found a tasteless white
⊖. He took 8 Lots of this ⊖ and 1 Lot of Miller's
☉, ground them until they foamed, poured a little
spirit over to make a kind of soup. Then he ground
it for another 2 or 3 hours, poured a lot of spirit
over to fully dissolve it and go through paper. Then
he filtered it and put it in a little flask,
extracted the spirit so that the remainder was like
an ⚇. He put this in a phial, left it to stand in
Balneo for 8 weeks, extracted the spirit again, and
was left with a yellow ⊖. He administered 3 grains
of this for all conditions.

I sent him the following process:

Oleum Vitriol,
3 Drops in all Circumstances.

A few days ago a knight called Salathurn
visited me and taught me a wonderful medicine made
from ♄. He took ♂ filings, poured 4 fingers of
sharp distilled vinegar over them, and put them in a

warm place, stirring every 2 or 3 days. He left them
until they had turned as red as blood. Then he
filtered the vinegar and extracted it until it was
dry. He ground the remains, poured them into an
unglazed bowl, sent it to a potter to be fired, then
poured over S.V. which had been extracted 3 times
from flowers of Sal Ammoniack. He used half a
measure of S.V. for 8 Lots of materia. He left it in
a warm place for 14 days and nights and the S.V.
extracted a good deal, leaving a brown earth behind.
Then he filtered the S.V. through filter paper and
saved it. Then he took 16 Lots of pure Roman
Vitriol, ground it to a powder, poured it into a
flask, and poured 1 measure of S.V.R. over it. Then
he sealed it, put it in a warm place and left it
until the ♑ had dissolved, except for a little
bit, then he filtered it, put it in a flask, and
gradually added drops of the ♂, solution, until 8
Lots had been added. He left it standing in a warm
place for 14 days and nights, well-sealed, and then
he filtered it. In the filter he found many red
feces. This is the corrosive ♀ ♂. He poured the
solution into a glass retort in the sand: applied a
gentle △, until the S.V. had all steamed off; then,
when no more clear drops fell, he took off the
receiver, emptied it of S.V., put it back on,
increased the △, and a blood-red ♋ evaporated

which is as sweet as sugar. He administers 3 drops of this in all circumstances.

Fauermann gave me some information in a letter: You have to grind the ☿ with the regulus in the warmth until almost everything has become ☿, looks tender and can be pressed through leather.

I communicated the following particular to him which an alchemist told me about:

Particulare ex Antimony & Min. Solis
cum Menstruo ex Stercore Humano.

He wrote to me and told me that I should take 4 pounds of the mineral ♂, coarsely ground, and I should calcine it in a secret △, until it was grey in colour, almost as if it were the smoke of the ⊖. Then I should proceed in the same way with ☉ filings. I should grind these two, put them in a flask and pour the following menstruum over them: Take stercus hois, pour it in a wooden vat, cover it and leave it under a roof for 8 or 10 weeks to rot. Then it should be distilled through a flask and I would have a spirit, an ♂♂ and a volatile ⊖. Then I should rectify the spirit with the volatile ⊖, and separate the ♂♂ from it. I should burn the ☺, boil it, and I would find a ⊖. I should add this to the

210

spirit and pour this over the finely ground minerals, so that the spirit covered it by 3 fingers. It should be put in a warm place until a skin forms on top. This happens within 24 hours. The skin should be taken off with an iron spoon and saved. I should repeat this until no more skin is formed. This should be finally put in a glass flask, the spirit should be poured from the materia, filtered, poured over the skin and then extracted per alembicum. One part of the remaining ♂ should be added to 4 parts ☽ in flux and it is refined into the finest ☉. It is made without corrosive. I haven't tried this myself, for I have only just been told about it.

Compendium

Referring to the Chemical Processes Described Until Now.

When I, Alexius Baron von Ruesenstein, came home from Salzburg, I put all my papers in order, as you will find here. I took the best of all of them and tried them out and found them to be true. So I'll tell you about them row, exactly as I did them:

Universal Tincture ex Terra Cemetern.

The very first thing was this: I began to examine the earth as Schulz had advised me to do. I got I measures of the very best fatty earth in which many corpses had rotted. I boiled it and poured the ∇ through a cloth, so that the coarse earth was left in the cloth. I coagulated the ∇ that had run through until it had reduced by a half. Then I filtered it through paper to clear it. Then I coagulated it gently in a bath until it was dry. Then I dissolved it in distilled rainwater, filtered it and coagulated it again. I did this three times. When it was coagulated dry, I put it in a little flask and poured a measure of S.V. over 16 Lots of the gum, sealed the top of the flask and put it in a warm place, but where it wouldn't boil. I left it until it was completely red, then filtered the S.V. through paper and distilled it per Balneum using a condenser. I poured the extracted ▽ over the ☉ in the flask and extracted it again twice. I did this 12 times. The extract was distilled per Alembic until it was dry. I was left with a red gum which I dissolved again in ▽, filtered and coagulated it gently. When I had filtered it, lots of gum was left in the filter, and the solution was a little clearer. When it was coagulated, I put it in a flask again, poured the S. V. over it again, and extracted it once more. I did this as often as before, until the S.V. was no longer coloured and was as clear as stream water on the maa. When this happened, I filtered it for the last time, and distilled it off gently, and I found a snow-white ⊖. Now the menstruum, as Schulz taught me, is ready. You must take care that the earth is not boiled dry all at once; it must be done 8 or 9 times, until the ⊖ comes out clean, for there is very little ⊖. I took 8 Lots of this white ⊖ and 1 Lot of Miller's ☉. I ground them together in a glass mortar until the materia was completely black and had almost become an ⚇.[19] This happened within 14 hours. Then I poured half a pint of S.V.R. over it, ground it for an

---

[19] The ⊖ becomes an ⚇ when it comes into contact with air. Hans Nintzel.

213

hour, then filtered the S.V. through paper, and it turned as yellow as ♃. This S.V. I put in a flask and distilled it per Balneum until it was oily.[20]

I was left with a red ♋ which I poured into a little phial.[21] I sealed it well on top with Spanish wax and put it in a Lamp oven, and heated it for 8 weeks with 1 light lit. The S.V. circulated, rising and falling, again and again. You could also see the most wonderful colours in the glass. After 8 weeks I lit a second light, and proceeded in the same way for 8 weeks. The materia became as black as ink. I waited until it lost its blackness. Then I lit the third light, did the same for 8 weeks and let it burn. In the eighth week the materia became red and some red precipitate could be seen at the bottom. The S.V. finally rebelled, and when this happens you should extinguish the lights and let the phial cool down. For the materia, when it sets, is fixed. Remove the phial and in it you will find clear spirit and red materia. Put a condenser over your phial and evaporate the S.V. until the materia is dry. Then extract the materia from the phial. Take 8 Lots of it and 1 Lot of Miller's ☉, grind them together and put them in a well-sealed phial in the sand. Apply the third degree of △ and leave it to

---

[20] This means reducing it by more than half. Hans Nintzel.
[21] The phial must be large enough for half to remain empty. Hans Nintzel.

stand for 24 hours. Let it cool, remove your phial and inside you will find a yellow lump. Break open the phial, grind the lump to a ♂⃰, then tinge. It will tinge 1 part to 10,000 parts. I have worked out a way of augmenting it: I took 8 Lots of the ⊖ which was as white as snow and added 1 Lot of ☉, as before. I ground them, dissolved them with S.V., filtered and distilled the mixture with a condenser; then I took 8 Lots of the yellow matter which was left behind and added 1 Lot of the tincture. I ground them and put them in a well-sealed phial, and applied the third degree of △ as before.

I let the phial cool down, and when I removed it I found a red massam. I took 8 Lots of it and ground it with 1 Lot of Miller's ☉, put it in a well-sealed phial, applied the third degree of △, leaving it there for 24 hours, and then I let it cool down. Finally, I ground it finely and could tinge 20,000 parts with it. I found this to be true.

Tincture Made from the Philosophical Water or May Dew.

Another time I worked with philosophical ▽, but I didn't follow any instructions; I made it up myself. This is what I did: I took as much ▽[22] as I could get, poured it into a kettle and smoked it until it was a liquor and looked a bit like a puddle of manure. Then I poured it into a barrel made of oak wood, put it under a roof or other airy place where the sun doesn't shire, covered it and left it until I could see that the ∇ was putrid and that little sponges were growing in the barrel or keg. I poured it into a flask, attached a condenser and receiver, well-sealed with luto. I put it in the sand and a spirit and a volatile ⊖ evaporated. When I saw that no more spirit was rising I took away the receiver. You know this by the following; as long as the spirit is evaporating, little trickles can be seen in the condenser, but as soon as it has all gone, the condenser dries. Save the spirit in a well-sealed container, but put the ☉ in a sugar grass and evaporate it gently until it is dry. When it is dry, put it in a flask and pour your spirit together with the volatile ⊖ over it. Seal your flask at the top with wax and put the flask in a gentle warmth for 24 hours, and the spirit will attract the fixed ⊖. Then filter it through wet filter paper and Flores will remain in the paper. That which is pure flows through. Then take 2 Lots of Miller's ☉, put it in a flask and pour 12 Lots

---

22 The water to which the author refers is the morning dew collected in May. -pnw

of your spirit with the fixed ⊖ over. Seal the flask well at the top and put it in a warm place for 24 hours. The spirit will extract the soul of the ☉ and will look yellow-red. Then filter your extract and the corpus of the ☉ will be in the filter, very black. This is useless; it cannot ever be reduced ad Corpus[23]. But make sure, when you filter your extract, that you first sprinkle what you are going to filter with spirit or S.V., otherwise too much a.a. sticks to the paper. When it is filtered, pour the filtrate into a flask and distill off the spirit until it is an oil. It should be like olive oil. When you have done this, pour the oil from the flask into a phial. Seal the phial at the top with Spanish wax over a stopper and tie it with a blister. Put it in a Lamp oven, light two lamps and let them burn until you can see through the panes of the oven that the maa., is as black as pitch. Before and afterwards you will see lots of lovely colours in the tube of the phial. Then light a third lamp and wait until it turns white and as red as blood. As soon as it is red, light a fourth lamp and let it burn for 14 days and nights. Then remove your phial and you will find a red, thick liquor. Empty half of your phial, attach a condenser to it, put the phial in the sand and extract the spirit until the maa.,

---

[23] Literally, to the body. -pnw

217

is quite dry. Then remover the phial, separate the bowls in the middle with a sulphur-candle and draw out your red massam. Take 8 Lots of it and add it to 1 Lot of Miller's ☉. Put this in a clear mortar and grind for a good hour. Then put it in a little glazed flask, well-sealed with good luto, and put it in the sand so that you can apply the third degree of △. Leave it in the sand for 6 hours, remove the flask and you will find a spongy lump inside. I ground this to a ♂ and tinged 10,000 parts of all metals. I also made a barren apple tree in my garden fertile with the liquor. I wanted to augment it, but I didn't have time. Instead I proceeded with other things out of curiosity. Now I will tell you the best way to augment both the quantity and quality, as Schulz told me in confidence: When you have prepared your tincture as described above, take 1 Lot of it and 8 Lots of the following maa. Grind the tincture finely in a mortar as described before. Pour the mixture into a phial which is so large that this maa. only fills one part of it, and three parts remain empty. Put it, as before, well-sealed, in a Lamp oven. Light the lamps exactly as before until the redness appears. Then ferment as before and it will tinge 20,000 parts and your tincture's quality and quantity has been augmented.

This is the materia which you need for the augmentation: Take the putrefied philosophical ▽ as before and extract the spirit. Also remove the phlegm from the ▽ and pour the spirit over the dried matter, from which you extracted the spirit and ▽ phlegm. Put it in a flask in a warm place for 24 hours, so that the spirit extracts the fixed ⊖. Then filter it through filter paper which is wet. Then boil off the spirit via the condenser until it is oily. Take 8 Lots of this oil, 1 Lot of your tincture and if you proceed as described before, that is how it is done.

Tincture ex Terra Philos.

Thirdly I worked with Schulz's secret tincture, or with the work cum terra sola which we learned first, but I didn't finish the work, because it was so much trouble for me. But I have seen that it works. I didn't ferment it, I just brought it to the red stage. I made it in the following way: I took 1 pound of ♂⚹, and the same amount of ♂↗, I poured it through 2 parts ♂̊. When the King had sunk, I ground it again and worked out the following process in my head. I melted the clinkers in the crucible again, and when they had melted I added the ground regulus.

I left it to melt for half an hour, then poured it out and proceeded in the same way as before. I did this seven times, and the seventh time I took the regulus, which had the consistency of melted wax, cleared it of its clinkers, pounded it coarsely and made an amalgam out of it with 2 parts ☿. I ground this until I could see that the regulus wanted to separate from the ☿. Then I dried the amalgam and distilled off the ☿ per Retort.[24] You will know when the regulus warts to part by the following; the amalgam gets very thick, looks quite wild, just like ♄, and the ▽ is the colour of ashes. When this is the case it must not be added to other ▽ because there is a lot of regulus in it. Instead it must be left to settle and the sediment must be added to the amalgam in the retorts. When all the ☿ has evaporated, take the ☉, melt it together without a flux and a vitrum will appear over the regulus. Put this aside, it contains nothing of use, but amalgamate the regulus again as before. Do this 7 times. The ☿ will happily separate or the 4th., 5th., 6th., and 7th. times, and it will give no sign of parting. Grind each amalgam for 24 hours, then distill the ☿ off. The seventh time leave it

---

[24] The amalgam must always be ground with warm ▽, as soon as the ▽ is dulled, another must be poured over. Hans Nintzel.

together, and grind your amalgam in warm ▽ until the ▽ disappears when you pour it over. Then dry your amalgam and put 7 or 8 Lots in a phial, place it in an oven, 2 fingers above the cupel, in the sand. Cover the bowl of the phial well with sand. In 4 weeks it will turn black, but it is over in 8 weeks.[25] When the blackness had gone I increased the heat of the △ and it turned the colour of ashes, not white, and 2 days later it was as red as blood. I did not carry on this work because no crystals formed as I had been taught. I should have proceeded as follows, as Schulz told me: I should have precipitated the regulus with ♄ and ☽ from the clinkers. Then I might have got the crystals from the field together. After putrefaction crystals may have formed which I would have imbibed with ☿ ⚖. Only then would it have turned red. But I didn't see any crystals because within 24 hours it had changed from black to grey, and within another 24 hours it was red. It could be that the other regulus, precipitated from the clinkers, turns the ☿ into crystals. The one I made does not do this. But it was a noble medicine. I tried it or a woman in Muhlstaten, not far from my castle, who lay terminally ill, I gave her one grain of my medicine.

---

[25] At first the second degree of △ must be applied. Hans Nintzel.

Three days later she visited me to thank me; she was fresh and healthy. If I had continued the process to the end, I should not have proceeded as Schulz told me, but as follows: I should have imbibed the red ♂ with 2 parts ☿ ⇌ as described before, and I should have let it run through all the colours. I should have done this 3 times. The third time I would have fermented it and I think that it would certainly have tinged 20,000 parts.

If you want to make it, I advise you to do it this way, for the method Shultz taught me is very dangerous and more troublesome. I should not have prepared the ☿ in this way either. I should have done it my way with 2 parts regulus of ♂. I should have amalgamated it 8 or 9 times, distilling it off each time, and it would have been purified enough. If you proceed as I tell you, you will achieve it without sorrow.

Tinctura Gualdi.

Fourthly I tried Gualdus's method which he sent me from Venice, together with a picture. But I didn't follow his instructions exactly, so I worked out a slightly better way: I did it like this: I took the same substances as in the previous process, smelted them and poured them through 2 parts crude

♂. When the King had sunk to the bottom I should have poured it through another 2 parts crude ♂, according to Gualdus, but I didn't. Instead I ground it with the clinkers with which it was first poured. Then I melted it for 1 1/2 hours. Then I poured it out, ground it and did it again. I did this 7 times, instead of pouring the King 7 times through fresh ♂ . My King stayed at a constant weight. Then I took 8 Lots of this King, melted it with 1 Lot of ☉, poured it out, amalgamated the mixture in the usual way with 3 parts ♀ revivified (as will follow at the end of this process) and then I ground the amalgam for 24 hours with warm vinegar. Then I poured off the vinegar, washed the amalgam with fresh ▽ until it was clear and dried it over a gentle glow. When it was dry I filled a little retort with it, distilled off the ♀ and melted the ☉ from the retort with the King again. I amalgamated it again as before with this distilled ♀ and ground the amalgam again for 24 hours. I did this 7 times, then dried the amalgam very gently, and put 6 Lots of it in a phial, put the phial in the sand in the oven, 2 fingers distance from the cupel and heated it with the fire as described in the previous process, until it reached the third colour. When it was red I imbibed it with the ♀ rectified; 1 part red ♂ to 2

223

parts ☿ like this: I broke open the phial and found a red ⚭ inside. When I poured it out I found a brown earth on the bottom of the phial. I put this to one side. But I ground the red ♂ in a stone mortar with the ☿. I continued while live ☿ could still be seen. Then I put 6 Lots in a phial again. I did this 3 times. The third time I fermented it with an eighth part ☉ like this: I took 8 Lots of this ♂, mixed 1 Lot of Miller's ☉ with it, ground it, put it in a clear smelting crucible and put it in a △ until it glowed brown. I left it to melt for half an hour and found a sponge like brown Corpus, the colour of ♃. I ground this to a fine ♂ and tinged 30,000 parts of all metals into the finest ☉ with it. The ☿ is prepared as follows: I take 1 pound of ☿ and dissolve it in 2 pounds of aqua fortis. Then I extract the aqua fortis until it is dry. Then I take the precipitate and mix with it 3 parts potash and 1 part iron filings. I distill it off per Retort in an open △ and the living ☿ evaporates. I used this for the first and the second operation. I found that this method worked. But I did not search further or continue my work with it. The other way could also be right. I can only be sure that this is what I found. You might be asking yourself why I have made so many tinctures. The answer is that I made them purely out of curiosity. I never tested more than one Lot.

Tinctura ex Stercore Humano.

For the fifth time I worked on Theophrastus's two methods: I found both to be correct. The first one was this: I took 1 pound of ☿ viva, dissolved it in 2 pounds of aqua fortis. When it was dissolved, I precipitated it with common ⊖ until a ♂ sank to the bottom. It is an advantage if the ⊖ is added to the aqua fortis in advance; if the ☿ doesn't precipitate, you have to add as much ordinary ▽, as there is aqua fortis; then it will easily sink to the bottom. When the ☿ had precipitated, I filtered the ▽, and the ☿ remained in the filter. You should wash out this white precipitate with warm ▽ repeatedly until there is no more acidity in it. Then dry it gently and save it. Then take as much Terra as you wish, let it dry out in the air or other warm place, then burn it to ashes. Leach out these ashes, filter the ▽ and coagulate it gently, and you will find a white ⊖. Dissolve this ⊖ again, filter and coagulate it. I did this 3 times. The third time, when it had been coagulated, I took

1 1/2 pounds of it.[26] I ground it with 8 Lots of the precipitate from before, put the mixture in a flask and sublimated it in the normal way. Then I put the sublimate on the bottom again with that which had remained fixed and sublimated it off again. I did this 3 times. The third time I took the sublimate as well as what had remained fixed, ground them together, put the mixture in a sugar glass in a cellar in wet sand, and it turned into an ⚭. When this had happened (it takes 14 days) I took pure calx auri or 1 Lot of Miller's ☉, poured 6 Lots of this ⚭ over it, mixed it in a glass mortar and ground it until it foamed and until the materia on the bottom was black. This takes 24 hours. Then I poured a rectified spiritum pluviae or spirit of May dew over it, but only enough to dissolve the maa; e.g. 8 Lots materia and half a measure of spirit. I ground it in the mortar for a good hour, then filtered it through paper and it was a nice yellow colour. The stuff left in the filter is useless, but I put what goes through the filter in a little flask, boil it off using a condenser until it is oily. I pour the oily substance into a phial which must be large enough that 3 parts are still empty, seal the top well, and put it in a Lamp oven. I proceed exactly as explained above, and I make a

---

[26] You have to burn a lot of earth to accumulate 1 ½ pounds of salt! Hans Nintzel.

tincture for 20,000 parts. I have found this to be
true. I haven't actually carried out the work as
Schulz told me. His could be right too. If you want
to carry it out, and you have gotten this far and
you want to augment its quality and quantity, then
follow my advice and you will not fail: Take 1 part
of your tincture after it is fermented and put it in
a phial with 8 Lots of this ♒ which has first been
filtered through paper to purify it. Put the phial
in the Lamp oven until the third colour has
appeared. Then proceed as before. You can do this as
often as you like. Each time it will tinge 10,000
parts more. You might be asking yourself if it is
possible that the ♒ could strengthen the tincture?
I tell you that the ☉ is unlocked during
putrefaction, and the better it is opened, the more
it will tinge. However often you augment it, you
have to first ferment it so that the tincture is
endowed with the fixed matter, and the menstruum
which is added is animated.

Another Tincture.

I found another way. I did the following: I
took 1 pound of ☿ viva and dissolved it in 2 pounds
of ♃ oil without adding ♁. This doesn't like

attacking the ☿, but if it is heated so that the ☍
begins to boil, then it will dissolve. I then poured
the solution into a phial which must remain 3 parts
empty. I sealed it well and put it in the sand where
it was heated, but did not boil. I left it there for
14 days. After a while I removed the phial, poured
the materia which was as white as milk into a sugar
glass, added ▽, and the materia turned quite
yellow. I waited until the materia settled, then
poured off the clear ▽, but then added more ▽,
stirred the maa. with a wooden stick and heated it
gently. This is so that the ▽ gets more power to
attack and wash out the materia. I washed it in this
way 4 times. The fourth time I dried my precipitate
over a gentle glow, then took 1 Lot of it and added
it to 2 parts ☿ rectified, (as will be described at
the end) in a warm stone mortar, and ground it until
the living ☿ had all disappeared. The liquid ☿ does
not have to be added all at once, but bit by bit,
and the mortar must be warmed up often. As you grind
you need the strength to grind it continually. I put
this ♂ in a phial which was a third empty and
sealed at the top with a stopper so that no dust
could get inside. I put it in the sand, 2 fingers
above the cupel, covered the bowl completely with
sand, and applied the second degree of △. In

8 weeks it was as black as coal, and in 8 weeks after that it was blood red.

N.B. All these methods using ☿ do not turn white. They go straight from black to red. When I write that it should go through all the colours, it is not true; various colours can be seen in the tube; this is what is usually called the peacock's tail, but the maa. turns only black, white and red. Natural matter, like May dew and the like, if it is made ex ☿, will only turn black, then red. But this means nothing.

When it was red, I took 1 part of this red ♂ and added 2 parts ☿ ⊒ at the beginning. I did this 3 times. The second and third times it turned red more quickly than the first time. The third time I fermented it with an eighth part calx auri in this way: I took 8 Lots of my red ♂ and ground it with 1 Lot of calx auri in a glass mortar. Then I put it in a smelting crucible, cemented it for 4 hours in a circulating △, very gently. Then I moved the △ so close that for a quarter of an hour the crucible glowed brown. Then I let it cool down and found a sulphur-coloured compact ♂ in it. I ground it on a grinding stone or in a glass mortar and tinged 10,000 parts of all metals. I have found this to be true. The ☿ for this method is prepared as follows:

I took 1 pound of ☿, dissolved it in 2 pounds aqua fortis. When it had dissolved, I distilled off the aqua fortis again. Then I applied a fierce △ so that all the corrosives evaporated. Then I let the retort cool down and poured the aqua fortis over it again. I did this three times. The third time I mixed the precipitate which is as red as blood with 2 parts potash. Then I added 1 part iron filings and boiled it off per Retort and the living ☿ evaporated. I needed this for the first amalgamation and for imbibing. I have found this tincture to be a true one, just as I have described it, you can rely or that. I am not saying that the others in other books are not true. I am just saying I definitely know this is true because I have made it with my own hands. You will have to decide for yourself. Now I will tell you which particulars I know of. I have collected them from many books, and have found them to be true. I am telling you the certain truth.

Particulare ex Copper, Iron & Gold.

First I will tell you my own method which I have, until now, only described vaguely. I will tell it properly and clearly, so that you can rely or it. I took 8 pounds of ♀ mineral and 8 pounds too of

iron and the same amount of ☉. I made it into a ♂⚹,
put it into a low receiver, poured 8 measures of
spirit of ⊡ over it, sealed the receiver at the top
with a stopper and blisters, and buried it in warm
horse dung where I left it for 8 weeks. Then I took
it out, poured the maa., which was as black as ink,
into several sugar glasses, put them in warm ashes
and let the spirit smoke off gently. When the maa.
was dry, I poured it into a retort, which must be
sealed with good luto and must be large enough for
half to remain empty. Then I put a rather large
receiver on it and began to heat it with gentle △,
then with a somewhat fiercer △. At first a sharp
spirit evaporated, then a white smoke as if I was
distilling a ♁. I continued with the stronger △
for 24 hours, then let the retort cool down. I
removed and found in the middle of the maa. a good
pint of spirit. I kept this in a well-sealed glass,
took 1 pound of arsenic album, 1 pound of
auripigmentum and one pound of ☿ ♎ sublimated with
♁ and ⊖. I ground and mixed these well on a
grinding stone, put the mixture in a glass flask
which could hold 5 measures, and gradually poured a
quarter of spirit over it. Then I sealed the flask
and buried it in horse dung as before. I left it for
8 weeks, took it out, poured the maa. into a very
large retort, large enough for half to remain empty,

which was covered with good lime. Then I put it in the △, attached the largest possible receiver to it, sealed it with egg white and chalk. Let the luto dry properly before you apply △. Apply the first degree of △ first for 4 hours, so that the retort becomes lukewarm, then apply the second degree for 8 hours so that the materia boils in the retort. Then apply the third degree for 24 hours. Take care that you do not go anywhere near the retort or the receiver, for it could cost you your life; there is nothing more poisonous or more penetrating than this, and you could not be saved from it. As soon as you have thrown coals into the oven, move away, for you will not have been able to seal it so well that nothing escapes. When the 24 hours are over, apply a wood △ for another 8 hours. This is the fourth degree. Then let it cool down. The luto on the retort must be a very good one, so that when the retort melts, the luto remains intact and does not let the spirit out. Make the luto like this: Take 8 pounds of good potters Tachet[27] and mix 2 pounds of iron rust with it. Cover the retort with this very thickly. In the fire it will turn to stone and no spirit will be able to penetrate it. Then take the receiver off the retort and pour the spirit

---

[27] Earth? Hans Nintzel. The Romanian-English translation is "peg", but potter's clay seems more correct here. -pnw

into a phial, large enough to still be 3 parts empty. Something will have sublimated in the neck of the retort. Add this too. Put the phial in the sand, 2 fingers from the cupel. Cover it with sand to the height of the maa. The phial must have a tube which is 3 spans long, and must also be sealed well with a stopper. The bowl of the phial must be covered with the same luto used for the retort. First the first degree of △ is applied for 4 hours to warm up the phial, then the second degree for 8 days and nights. Every six hours the stopper must be pulled out, because the materia is boiling, and through the boiling forms a wind which could cause the phial to explode if it were not released. When the time has passed, let the phial cool down and pour the refining ♒ into a flask, or leave it in the phial. Add old money or ☽ lamellas, or if you have no silver, add some drops of ☿. In 14 hours it will have become clear ☉. You can do this again and again until you have no ▽ left. It needs no fermenting like other refining ▽. When the metal is in it, put it in a half-lit place and leave it to stand for the time it needs. Then remove the lamellas and they will be as black as coal. Smelt them together and you will find the finest ☉. With 1 pound of this ♒ you can turn 32 pounds of ☽ or ☿

into ☉. This is true. I have done it for many years and you can rely on it.

Now I will tell you about the particulars which I have tried and found to be true.

Particulare ex Mercury Triturando.

First I made a fixed ☿ ii. I have always known how to make one of these, but my method was not as good as this one: I took 1 pound of ♁ and ♂, mixed and smelted them and poured them through 2 parts crude ♁. The King, which sank to the bottom, I melted in a crucible. Then I added as much glowing ♂ as I wished and kept adding crumbs of ⦻. I left it standing in a strong △ until I could see that the regulus was absorbing some of the ♂. Then I poured it into a Giesspuckel[28], and let it cool down in there. I ground this regulus coarsely, so that the pieces were the size of peas and amalgamated it with 2 parts ☿ viva, as I told you before. I ground the amalgam finely by hand, the put it into a triturating mill and bricked the mortar into an oven, under which a constant △ was burning, so that

---

[28] A Giesspuckel is a conical shape of brass or iron, into which the molten metal is poured to quickly solidify it. -pnw

the ▽ boiled constantly. I ground it for 4 weeks. Every 12 hours I poured away the ▽ and added fresh water. Nearly all the ♂ became ☿. Then I pressed the amalgam through leather, I used what went through the leather for my work or fixation. The rest of the amalgam which didn't go through the leather I saved. Then I took 4 Lots of fine ☽, dissolved it in 16 Lots of aqua fortis. When the ☽ was dissolved, I added as much ▽ to the aqua fortis as there was aqua fortis, and precipitated it with common ⊖. When I had done this I filtered the ▽ and I was left with the ☽ in the filter. I washed this well with warm water, dried it, put it in an iron mortar, poured 1 pound of the ☿ which went through the leather over it, heated the mortar and ground it with great force. An amalgam was formed. I then poured warm ▽ over it and ground it for 24 hours. I replaced the ▽ with fresh water when it became dull. Then I dried the amalgam and put 16 Lots in a phial which I put in a fixing oven in the sand. I covered the bowl of the phial with sand, 2 knife blades above the cupel. Then I applied the second degree of △ for 6 days, then the third. I proceeded like this for 6 weeks. Then I let it cool and poured the red precipitate which would come out easily from the phial. The compact substance

235

that was left I filed finely. I took 1 part of it and added 4 parts of ☿ made in the way I told you about before. Do the following: Take the red ♂ and the filed corpus, and for every pound of this mixture, add 4 pounds of the precipitated ♀. Grind them together in a mortar until you have made an amalgam. Then melt it as before. Do this 4 times. The first time, half will remain in the cupel, the second time a third will remain, and the third time all will remain. The fourth time you will find clear gold. The first time the seventh part separates. The second time it gives a fourth part. The third time it is 3 parts ☉ and 1 part ☽. The fifth time it is too highly coloured and is useless for a ferment and you have to make a fresh one. When you have done everything as I have told you, distill it with ♄ and you will find what I said you would.

Another Particular.

I have also made the following particular: I took 1 pound of ♀, and dissolved it in 2 pounds of aqua fortis. Then I also dissolved 16 Lots of ☽ in aqua fortis. When both had dissolved, I mixed the two solutions and added the same amount of ordinary

∇ to them as there had been aqua fortis. I also added common ⊖. Then I filtered the ∇, and my ☽ and ☿ stayed in the filter. I washed these two with warm ∇, dried them very gently and saved them for future use. Then I did the following: I took 1 pound of ☉ and ♄ a.a., and burned them, ground them finely and mixed 3 parts potash with them. I put everything in a kiln and fired it. Then I boiled it up, filtered the ∇, coagulated it in a sugar glass, and, when it was dry, I put 16 Lots in a glass, put the glass in the fresh air, and it became an ⚬⚬. Then I took 4 Lots of the precipitate, ground it finely and put it in a small earthenware flask. I poured the oil over it, attached a condenser, put the flask in the sand, and boiled off the moisture until it was quite dry. The flask must be covered in the Luto I told you about before. When it was dry I increased the △, and the ⊖ melted. When I thought it had melted I removed the condenser. If some of the ☿ had condensed in the condenser I put it back into the flask. I put the flask in the fresh air. Overnight it turned into ⚬⚬. Then I proceeded as before. I did this 7 times. The seventh time, when it turned into ⚬⚬, I filtered the oil through paper. I coagulated what had flowed through the paper gently. When this was done I added 4 Lots of this to 1 pound of warm ☿ and it fixed to ☽. This is true.

Another Work.

I have also worked with ♄, like this: I took a large measure of potash and two measures of beech ashes. I boiled them, and made a very sharp lye from them which I boiled. Then I added more lye. Then I took de igne nostro[29], ground it to a subtle ♂ and put 4 pounds of the same in a kettle. I poured over 2 quarters of the lye and let it boil for 6 hours. Then I boiled it until it was like a liquor. I poured this liquor into a glass, covered the top with paper and put it in the cellar for 4 weeks. Then I poured my maa. into a pan or kettle, poured a quarter of common water over it and boiled it again for a good hour. Then I filtered the ▽ and washed out the substance which was left in the filter with warm ▽ until no sharpness was left in it. Then I dried it over a very gentle glow. I had to make sure it didn't melt, because it is soft. When it was dry it was as red as blood and liquid, like wax. I place one pound of ♀ and 8 Lots of the fixed ♄ in a crucible in a circulating △ for 6 hours until the

---

[29] Literally: Our fire. -pnw

♃ flows over the ☿. Then I let it cool down. I break open the crucible, and I find a lovely coagulate; 24 Lots from one pound. I make the fixing flux for this coagulate: Take the lye with which you boiled your ♃ic earth, coagulate it gently to a ⊖ as taught in the previous process (after it is fired in the kiln, boiled and coagulated), mix these two ⊖ and add 8 Lots of your flux and 8 Lots of your coagulate. Grind them, put them in a crucible, cement for 4 hours in a circulating △, then smelt it all together. You will get 24 Lots of ☽ from one pound of ☿. In the core you will find 3 Lots of ☉. This is true.

Another Particular.

I have also made a true augmentum which I have already described to you, but I would like to show you it a bit more clearly. I took 1 pound of ☿ viva, and amalgamated it with 16 Lots of regulus ♂. I ground the amalgam. Then I ground it further in a triturating mill until it all looked like ☿. This takes 4 weeks. Then I pressed this amalgam through leather. I took 16 Lots of it and made it into an amalgam with 2 Lots of calx auri. I did this in a mortar with a lid to prevent any dust from escaping.

I ground it until it turned into a black ♂. Then I put 8 Lots of this ♂ in a phial which I sealed and put into the sand, 2 knife blades from the cupel. I covered the bowl of the phial with sand and applied the second degree of △ until the contents of the phial were as red as blood. Then I let the △ cool down, removed the phial and took 8 Lots of the ♂ and added 16 Lots of the previous ☿. I ground it again as before until I couldn't see any more ☿, and I melted this ♂ again as before. You can do this three times, but you must augment it before you boil it off. The third time the weight of the ☉ remains the same. The 4th. and 5th. time it tinges the ♄ on the cupel. Finally it can tinge so high that it is inconceivable. I have done it 300 times and have made 150 pounds of ☉ with it. I recommend it to you, for it is true.

A Particular or Refining Powder.

Furthermore, I made a refining ♂ as follows: I dissolved 1 pound of ☿ in 2 pounds of aqua fortis, then added the same amount of ▽, and precipitated it with common ⊖. Then I filtered the ▽ and washed out the precipitate with warm ▽, until all the sharpness was gone. Then I dried it gently, mixed it with twice the quantity of flowers of Sal ammoniack made with ⊖ ground it, put it in a flask and sublimated it in the normal way. I let it cool down, took 8 Lots of the sublimate, added 4 Lots of lamellas ♂, left it in a sugar glass in a cellar and within 24 hours it had melted to an ♋. I filtered the ♋ through paper, poured it into a flask and added 1 Lot of calx auri for every 8 Lots of sublimate. I extracted the water until it was dry, then poured it back over again. I did this 12 times until the ♂ was as red as blood. Then I took it out of the flask, ground it finely and added 1 Lot of this to 8 Lots of ☽ in flux and it was refined. If you want it to refine more easily, cement the ☽ first 4 times with potash and ⊖ to make it brittle.

I didn't try anything else with this. Memorize one of the other processes and perhaps you will find something else, there are such lovely things in these books. If I were to try them all it would take about 20 years. But you can rely or these processes which I am telling you. If you have no means to start working, sell one of these processes and you will be able to use the money you get for it to start your work.

How the Metals are Prepared for these Works.

The ☉ is poured through ♂ 7 times and purified of all foreign bodies, as gold-beaters do. Then it is beaten thin.

☽ is also purified 7 times and beaten to thin little plates.

☿ can be evaporated 7 times via a condenser or retort and it is good for amalgamating. If you want to increase the tincture it can be washed with vinegar and ⊖. Make sure you get virgin ♀ium because nothing else is mixed in with it. Many people pour it through gold casts, and it is easier dissolved. When it has been distilled 7 times it is quite black and powerless. Revive it with one part potash and the same amount of ♀ albus, and it will live again and be of the colour of a blue sky and clouds. Other people have other ways of purifying it.

The Second Journey

Extract from the Conversations of Baron von Ruesenstein and the Adepts on Another Journey.

The Round Root, a Universal Medicine.

These things are called round roots and they grow between stones on the shaded side where lots of moss grows too. They can be found in the middle of the moss. They have leaves like those of a carrot, but they are very small and you can hardly tell the difference between them and the moss. The root is as big as a hazelnut and formed just like a normal root. It is so precious that 3 grains of this root can help cure all illnesses.

Schulz once said: My tincture is not in the regulus, nor in another metal; it is in Nature herself, for it is in the May dew, rain or snow. For the nature of metals is fixed, but these are open and nature is hidden in them in a natural way. The ⊖ can be brought out through putrefaction or through the skill of our artists. This is both the beginning and the end of all things.

A Tincture for 14 Parts.

He ground 1 pound of arsenic finely, then set it aside, Then he detonated 2 pounds of crude ♀̄, and 2 pounds of ⊕ and mixed it with 3 parts potash. He fired it in a kiln. When it was burned well he boiled it in ▽, filtered it and coagulated it dry, mixed it again with 3 parts potash and burned it as before. He did this 3 times. The third time, when it was boiled he coagulated it in a sugar glass, when it was dry he added 1 pound of this ⊖ to 16 Lots of arsenic in a mortar and ground them. Then he put this mixture into a sugar glass in a damp cellar and left it until it had turned into an ⚯. Finally he poured the ⚯ into a well-coated flask, extracted the moisture and then applied a fierce heat so that the materia could go into flux. He left it in this fierce △ for 24 hours. Then let it cool, then removed the flask and broke it open. Inside he found a fixed white stone which he put back in the cellar to become a liquid again. He did this 7 times. The seventh time he fermented it with ☽ like this: He took 4 Lots of fine ☽ and dissolved it in aqua fortis. Then he added the same amount of ▽ to it, added ♀ plate and put it in a cool place. The ☽ sank white to the bottom. Then he filtered the ▽

and the ☽ remained in the filter. He washed this with warm ▽, to remove the sharpness and dried it over a gentle coal △. Then he took 16 Lots of his stone, added 2 Lots of calx auri, ground them together, put the mixture in a glazed flask, put it in the sand, lit a circulating △ in the normal way, then kept moving the △ nearer until the materia melted. Then he let the flask cool down, broke it open, removed the stone which was as white as snow, and used it to tinge ♀ and ☿ into ☽. One part makes 14 parts. This is certain.

A Particulare: One Part for 8 Parts.

I took 1 pound of Hungarian ♁ and poured one measure of S.V. over it. I left it standing in a warm place until the ♁ had almost completely dissolved, excepting a few feces. Then I filtered the S.V., boiled it over a little flask until it was dry, then poured the S.V. back over it and dissolved it again as before. This should be repeated until the ♁ quite white and as liquid as wax. This ⊖ is the true ⊖ of metals and needs nothing more than a ferment of ☽ or ☉. It will tinge according to the ratio of the ferment. I fermented it like this: I took 8 or 9 Lots of the liquid ⊖ and added 1 Lot of ☽ or ☉. I put it in a glazed and well-coated flask in a strong △ and left it for 24 hours until the ⊖ melted and absorbed the ferment instead of its crude parts. After this it tinged 9 parts of ♀ into ☽ when it was fermented with ☽. If it is fermented with ☉, 1 part tinges 8 of ☽ into ☉.

Once, when Herr Schulz and I were on a journey to Venice to visit Gualdus, we were discussing, amongst other things, Basilius's ♁. We were saying that this must be no common vitriol, but one from ☽, or ☉ mines. I said: Dear Schulz, does ☽ also contain a ♁? He said; yes, have you ever seen a single metal or earth which has no ♁? There is nothing on earth which is not joined to ♁. ♁ is the first principle from which the metal springs, for it is a spiritual seed formed by the air and the ▽ in the earth and which is heated with the help of the △, or sun. I said: But, dear brother, do we build more on the ♁ of the metals in our art, or on the minerals and metals? He gave me the following answer: Dear brother, ask Albertus Magnus who says clearly in his third tract: With the beginning make the end. Take the end and make with it a beginning, and then an end again. You would fail badly if you were to make a beginning from an end, for it would take longer to make a beginning from an end than it would to make an end from a beginning. Therefore take a beginning and make it into an end. But which is the beginning? The beginning of all metals and all things is a ⊕~. What is a ⊕~? A ⊕~ is a ⊖, a solvent which dissolves all waters. For this reason he also writes that the materia for this art is to be found in all things. He says there is nothing in which the true maa. is not hidden and from which it can easily be extracted. I answered: This is true, I have read this. But I cannot find my way into this, and I do not know how to, because you, dear brother, and all the people I have corresponded with just shout and write that the philosopher's stone is made ex ☿ & cum ♀. They also write that the true materia can be found in ♄ and in ♂, as Monte Schider clearly states. Philaletha also says that the secret way lies in ♄ and in ☿, and your dear brother always say that there are only 3 ways to the Philosopher's stone; the secret way is ex ☿, solo & ♄, the middle way is cum ♄, ⊖ & ☿, and the hard and last way is cum ☉ & ☿. Theophrastus also found a way ex solo ☿. Who should I believe? I am going mad trying to find my way in all this. Schulz answered: Dear brother, I think you are a bit of a fool when you make such speeches, and I also think

you think l am a fool because I told you that the ♁,

as I prepare it with ☿ is the secret way. This is

true, but it is not the secret way of the ancients,

but of the new philosophers and chemists. It is true

that I have found this to be correct and have used

it to make the tincture, but it is not my most noble

work. I only tried it out of curiosity to see what

was in this famous ♁. So many people praise it that

they give it a most noble symbol; the earth's globe.

But it is true that there is a lot in the ♁, it

contains more than many minerals, as long as it is

still in its minera. But once it has been melted out

of its minera, it loses its strength. Why do people

say that the ♁, contains so much? Because it is an

open mineral, and, so to speak, a pure one, but not

🜔 which will dissolve in every ▽, but a 🜔,

which dissolves in △ and which destroys other

minerals and metals. Just like aqua fortis, which is

extracted from the 🜔 with the help of ⊖, and

which destroys metals and minerals in moisture, ♁

destroys them in △. When it is calcined in the

secret △, the finest 🜔 can be extracted from it,

as I have already told you. ♁ is a wolf or destroyer

of all metals in its unlocked and 🜔ic state, and

it is liquid at the same time. This is why it is given the symbol of the globe. Now, the materia is also in other minerals and metals, but it is something supernatural, like in arsenic, ♃, ♀, ♂, ☉, ♄, or in ☿ sofo. But the true way is ex ⊖ as I have already told you, and of which I will also tell you more. This is the true way of nature. I advise you to remain or this path.

Another time he said: Look, nature's sap can be found in the pine and oak trees, not just in these two, but also in all herbs, roots, horse dung, human excrement, grass, trees and similar, no matter what.

Menstruum ex Vegetabilibus & Stercoribus.

I take one of these two substances, put it in a flask or a retort, distill it in the usual way to remove all the moisture, and burn the ☉ on the bottom. What has evaporated. I pour into a wooden vessel, well-sealed, put it in the fresh air where it is warm (this must be done in the summer), and leave it for 3 months. It putrefies and through the putrefaction all things are separated: This is how the ⊖ is separated from the spirit and the phlegm, which is from the earth, sinks to the bottom. This separation causes a great stink and putrefaction. When it starts to stink, pour it all into one flask

249

and extract the spirit per alembicum. There was a man who did it using an oak tree, but he didn't take the wood. He took the foliage and the acorns. He pounded them and squeezed out the sap. He put the sap in a vessel, let it putrefy and then proceeded as I told you.

Gualdus once said: The tincture is nothing other than gold. But this is only the menstruum, a ⊖, which opens the ☉ and introduces it into unripe metals and ripens them at the same time. It is important for a man who works in our art to know that he should search for the true menstruum in one thing. It can be found in nothing other than the salted materia, which contains no ⊖, but which brings forth a ⊖ during putrefaction.

One of Gualdus's Particulars.

I took 4 pounds of ♂ and poured 4 measures of distilled vinegar over it. I did this in a glass bottle, which I put in a warm place and left it to stand until it was as red as blood. Then I filtered it through paper. Then I added drops of ⊖ ▽ until some ♃ sank to the bottom. The ⊖ ▽ must be made like this: I take 1 pound of common ⊖ and dissolve it in one measure of ▽. When the ⊖ has dissolved,

I filter it. I precipitate the ♁ from the vinegar with this ⊖ ▽. When the ♁ has been precipitated. I filter the ▽ and the pure ♁ is left in the filter. I wash this well with warm ▽ until all the sharpness has gone. Then I dry it gently on some paper. Then I take 1 part of it, 3 parts flores of Sal ammoniac and 1 Lot of Miller's ☉ for every 8 Lots of ♁, and grind these things together. I put them in a glass bowl in the cellar and leave them there until the materia is quite wet. Then I pour it into a flask and boil off the moisture, and then I sublimate it in the normal way. I pour what has evaporated into the condenser, wash it out, then pour it back into the flask and repeat the whole procedure again. I do this 15 times, and then it is done. If I take 1 Lot of this materia and add it to 1 pound of ☽, it is turned into the finest ☉.

Tinctura Universalis.

There was a man who took a Sublimate and put it in an earthen bowl, and had it fired by a potter. Then he let it melt to an ⚭. He put this ⚭ in a flask and boiled it dry. The ⊖ that was left behind he melted again, then boiled it off again as before. He repeated this until he had only a very little ⊖.

251

Then he took all the ▽, poured it into a large glass, sealed it well and placed it in the sun or other warm place. He left it there for 3 months. Then he took this putrefied ▽, poured it into a flask, boiled off the spirit, separated the phlegm too, and, when this was gone, he found a pure ⊖ in the glass which was white and liquid, like wax. He took 8 Lots of this and ground it with 1 Lot of Miller's ⊙ in a mortar until all the ⊙ had disappeared. Then he poured some spirit over it, ground it for a while longer so that the ⊖ could dissolve with the ⊙, filtered the materia through paper, poured it into a flask and boiled off the spirit again. He was left with a fine ⊖ which was a golden yellow colour. He used this to tinge, but it was no use. To bring it to the right path, he proceeded as follows: He did not boil all the spirit off the ⊖; he stopped when it had become oily. He poured this oil into a phial which was big enough for 3 parts to remain empty. He sealed it hermetically on top, put it in a Lamp oven and let it go through all the colours as normal. When it was red he fermented it with a ninth part ⊙ and tinged 20,000 parts of all metals.

Gualdus once said to Schulz: You rightly say that one herb is better than another. But it depends whether you want it for a ♂, a ▽ or an extract. But if I turn this herb around, and take the first beginning from it (which is the ⊖), then this ⊖ is like everything; vegetable, mineral or metal. But how can this be? Perhaps one has to burn the herb and leach out the ashes, then filter and coagulate it? No, the ⊖ must be extracted without burning; for if it is burned, the primum ens disappears, that is the spiritual ⊖ or the life of nature. It must be ripened by decaying; it must be made fixed so that it can penetrate human beings as well as metals, so that it can ripen raw metals, so that it can destroy evil in humans and take effect. The ⊖ of all things is life - in humans, in herbs, metals or minerals. These are all flammable things made by God, each with its own form.

Universale ex Water Maris Nostri
of Herr Fornegg.

I took some rainwater and kept it in a barrel on the ground, where the wind could blow through it, but the sun couldn't shine into it. I covered it and left it there for a whole summer. It decayed and the ▽ stank and was full of threadworms. When I saw this I poured it into a flask and distilled it per alembicum. When I had all the spirit, I smoked the remaining phlegm until it was dry. I poured the extracted spirit over this dry gum, sealed the glass at the top, put it in a gentle warmth and left it for 24 hours. Then I filtered it, poured it back into a flask, and extracted the spirit until it was dry. I found a white ⊖. I took 8 Lots of this and 2 Lots of Miller's ☉, and ground them together in a mortar until a black foam rose. Then I poured 16 Lots of my spirit over it and ground it for a while longer, filtered it finally through paper, poured it into a flask, and extracted the spirit again until it was dry. I found that the materia in the flask was quite yellow and was liquid, like wax. Then I took as much ☿ as I wished and boiled it in the spirit for 24 hours over a strong △. I poured spirit over until it, covered it by 3 fingers, and it was purified. Then I took 8 Lots of this purified ☿ and 4 Lots of the previous materia and mixed them together in a mortar with a stick. I made a nice amalgam which I put in a phial. I didn't put more than 6 or 7 Lots in the phial, which must also remain 3 parts empty. Then put it in a dry bath, not in the sand. For, dear brother, if you put it in the sand, the heat is fierce where it is covered with sand, and where it is above the sand, it is cold. But do you know what a dry bath is? It is just like a sand cupel without sand. Through the middle are two iron bars, on which the phials stand. The one on top is the same size as the one below. The cover has a hole to let out the heat. It must have a tube, covered with lime; otherwise the escaping heat will crack the tube. Outside the cupel, you must have the normal verts, so that the gradum ignus can be modified at all times. There must be a hole through the cupel, in which there is a pane of glass so that you can see in and watch the colours charging. Apply the first degree of △ for the first 6 weeks (i.e. open one vent) then gradually apply the second degree until it is blood-red. Finally, ferment it with the ninth part ☉ in a crucible and it will tinge 30,000 parts of all metals.

The Best Universal Method ex Water & Terra.

Our teacher Basilius took earth which had been locked up for three years and never opened. He took three measures of this earth, dried it in the open air, and when it was as dry as dust, he gradually poured it into a large kettle, not too much at a time, just a little, bit by bit, so that it could be boiled. He added ▽ aretis or ▽ common or ▽ gemi. (which must not yet be decayed) and boiled it well. Then he filtered the ▽, evaporated all but a very small quantity, so that only a quarter was left and three-quarters were steamed off, and it was quite red. He continued until the earth was boiled clear. Then, when he had all of this ▽ in a Compendio, he poured it into a wine barrel (from which the ♄ had been removed) and put the barrel under the roof, where it was warm and the sun couldn't shine, covered it with boards, and left it until the ▽ was full of worms and stank like a rotting carcass. Then he took the ▽, poured it into a still and distilled it as usual. When all the spirit had evaporated he let the remaining ▽ evaporate. Then he found a gum at the bottom which was red, like clotted blood. He poured the evaporated spirit over this (he called it spirit of ☿), sealed the top of the flask so that the spirit couldn't escape, and gently digested the flask for 24 hours. Then he filtered it and some feces were left in the filter. That which had filtered through, he poured into a flask, and then distilled off the spirit until it was dry, poured the spirit back over it again, and continued as before. He repeated this until the ⊖ was as white as snow. Then he took 1 Lot of Miller's ☉, 8 Lots of this ⊖, ground them together in a glass mortar until the ⊖ was as yellow as ☉, and until a ♂ could be seen at the bottom of the mortar, then he poured it all into a glass flask and put it in a gently heated place and left it for 24 hours. The spirit extracted some a.a., from the ☉ which the ⊖ could not extract. Then he filtered it through paper, poured it into a flask, extracted half of it, and poured it into a phial which remained 3 parts empty. He smelted the phial shut at the top, put it in the Lamp oven, just as Fornegg explained, fermented it with the ninth part ☉ again when it was at the red stage, and could tinge 40,000 parts of all metals. This is the best of all methods on earth. I recommend it to you.

Once I said to Schulz: Dear brother, Fornegg told me what he had withheld from me until row. He also told me that he knew for certain that one of you had kept part of the work from me. Schulz said: I know for certain that I haven't kept anything from you. Indeed I have revealed everything to you correctly. I assure you that you should do with the ♂ as I told you, but without the ⊖ it is hard work. For the ⊖ is added so that it can extract the a.a, and introduce it into the ☿. But if the ☿ is prepared properly, it chases the bad earth away, and only pure a.a. remains in the ☿, and turns into a tincture. For the ⊖ which is in the rain also contains the ♂. If it is dissolved properly, it is almost like the other. When the ⊖ is added to the ☉, the ♃ has a higher ☉ colour in it than the ☉ itself. If it didn't have a higher colour, it could not raise the colour of the ☉. I will repeat what the ⊖ can do. It is better for this work, it can tinge higher and more things, it is not as dangerous in the △ as when this is done without ⊖. This is what must be done: Make the ⊖ into a spirit from the regulus or tincture, as you have been taught. Make my ♃ic ☿. When it is properly ♃ic, grind it to a ♂, finely as you can, then take 4 Lots of this regulus and add 8 Lots of the regulus you already made. Grind them until you can see that they are nearly all foam. Then pour as much spirit over the mixture as is necessary to dissolve it, filter it and extract the ▽ via a condenser until it is dry. Then take one part of this yellow ⊖, 2 parts of ☿ rectified (dissolved with aqua fortis, precipitated with ⊖ and sublimated and revived), and put it all together in a mortar, grind it until it is an amalgam, put it in a phial, and you will have made what I told you, only somewhat quicker.

Oil of Vitriol,

which also belongs to Ruesenstein's Process.

I take 8 pounds of ⊕~ common and 4 pounds of mineram ♂. I pound them and mix them and distill the mixture per gradus △ in a retort. When all has evaporated, I grind up the ☉, pour the spirit back over and proceed as before. I do this three times. The third time, when the spirit has evaporated, I quickly pour it out, attach the receiver again, and I get a red sweet ⚭. I add this ⚭ to the Spiritu mineralium, and all the corrosives are precipitated, and sink as feces to the bottom. Then I filter it through cotton and do as I said before. The other ⚭ of ⊕~, described before is only for use in augmentation; when the ☽ is already as black as coal, I add this and it aids the other one; it helps it penetrate more easily and colours it at the same time.

After I had tried all the things which I have described in this Compendium, this was in the year 1663, Herr Schulz came and visited me on the 14th. of June, 1664. He stayed for 3 days, and was made welcome by myself and my wife. He told me what he had been doing since he last wrote to me from Salzburg. He also asked me what I had been working or since then. I answered: What I have written in the Compendium. Schulz asked me if I had searched in the regno vegetabili. I said no. He said: Dear brother, much is hidden in the vegetable kingdom. I tell you, there is a certain root which grows in the highest mountains. If you administer only 3 grains of this, you can cure a person of all illnesses. It is well-nigh a universal medicine. Then in order that Schulz might see that I already had something real, I said I would show him what I had made. I brought a fire and a crucible into the room, 3 pounds of ☿ and some ivory boxes. I tinged in three different ways before his very eyes. He asked me what sort of tinctures they were. I answered: One I have from Gualdus, one from you, made of May dew, and the third is made of ⊖ hois. I made these, together with 2 or 3 particulars from ☿. Schulz said: Much is hidden in arsenic too. When we get to Verine I will tell you much about arsenic.

Particulare 1 Lot ad 8 Lot Silver in Gold
ex Mercury, Iron, Sulphur, and Gold.

An alchemist sublimated 1 pound of Magesterium ☿ and 1 pound ♃ common together. Then he made a ♎~ of ♂ with ∞ of ♃ and extracted it with S.V.R. He

took 4 Lots of the extract and 8 Lots of the
sublimate, ground them, sublimated them until
everything was fixed (this takes 7 or 8 weeks), then
took 8 Lots of this and 1 pound of calx auri,
cemented it for 3 hours until the crucible glowed
brown, then added 1 Lot of this ♂ to 8 Lots of ☽
in flux. It was refined. This is true. Each of the 4
species has its own tincture.

A Tincture ex Salivae Menstruo.

There was a man who took saliva, let it putrefy
in a little barrel under a roof where the sun and
wind could get to it (but was sealed tightly), and
then he distilled a spirit and a ⊖ volatile. He
mixed these together, smoked off the ☺, poured the
spirit over it, sealed it and left it in a warm
place until it was dissolved and then filtered it.
He took 8 Lots of this, 1 Lot of ☉ and dissolved
it, filtered it again, so that the Terra ☉is was
left behind and extracted the spirit until it was
⚇y. He treated this ⚇ with the 2nd. Degree of △
until it was red, hanging in the Lamp oven. Then he
added 1 Lot of ☉ to it, left it for 24 hours in a
warm place and the materia dried. Then he applied
the third degree of △ to the phial for 6 hours. He

found a sulphur coloured ♂ which tinged 30,000 parts.

The Preparation of Salt from the

Earth Dug Up by Moles.

Take some limey earth which moles have dug up, leave it in a cellar for a whole year, extract the salt and coagulate it. Take 8 Lots of it and grind it with 1 Lot of ☉, pour spirit of rain over it, grind it again, filter it, put it in a phial and boil it dry, as Gualdo tells you. The earth should be opened up and sprinkled with rainwater every day.

There was a man who made a menstruum from rainwater and earth. He put them in a room with other fatty earth, sprinkled them every day with putrefied rainwater, and left this for the summer. Then he distilled it fiercely per retort. He put the water in the sun in sealed glass containers for 3 months. Then he distilled it. He had a spirit and a ⊖ volatile. He evaporated the ☉, poured over the spirit, dissolved and filtered it, boiled the remaining earth, filtered and coagulated it, and he was left with a small amount of ⊖. He took 8 Lots of this, 1 Lot of Miller's ☉, ground the two in a stone mortar until he could see no more ☉, poured as much spirit over it as there had been ⊖, ground it again for a good hour, filtered it, distilled it to an ⚯, boiled this as before, and he had made a tincture for 30,000 parts.

Finis.

A Word from the Publisher

Thank you for purchasing this book from The R.A.M.S. Library of Alchemy. During his lifetime, Hans Nintzel dedicated himself to the identification, acquisition, study, retyping and, when necessary, translation of what he considered to be the most important known works on Alchemy. Hans was assisted by his sparse network of fellow Alchemists, all members of the Restorers of Alchemical Manuscripts Society (R.A.M.S.). I was an active member of R.A.M.S.

Hans provided copies of the R.A.M.S. works as photocopies. My goal is to publish all of them as professionally printed books.

The works from the original R.A.M.S. Library are republished by R.A.M.S. Publishing Company in the collection, "The R.A.M.S. Library of Alchemy," with permission of the Estate of Hans W. Nintzel.

If you have a work on Alchemy that you believe should be a part of the R.A.M.S. Library, please contact me through R.A.M.S. Publishing Company.

Philip N. Wheeler

The R.A.M.S. Library of Alchemy

The study and practice of Alchemy was extremely important to Hans W. Nintzel. He assembled this Library over a period spanning more than three decades, guided by his teacher Frater Albertus. The R.A.M.S. Library of Alchemy includes all of the most valuable Alchemical texts that Hans painstakingly located, acquired, retyped, and translated during his lifetime, with help from other R.A.M.S. members.

The following is a list of the volumes that are currently available. Volumes that contain works from multiple authors may have only the principle author or editor listed. Additional volumes are forthcoming.

Volume	Title	Author or Editor
1	Twelve Keys of Basilius Valentinus	Basilius Valentinus
2	Triumphal Chariot of Antimony	Basilius Valentinus
3	His Secret Book	Artephius
4	The Golden Work	Hermes Trismegistus
5	Three Works of Ripley	George Ripley
6	Four Works of Paracelsus	Paracelsus
7	Bacstrom's Notebooks, Part 1	Sigismund Bacstrom
8	Bacstrom's Notebooks, Part 2	Sigismund Bacstrom
9	Summa Perfectionis	Geber (Abu Musa Jabir ibn Hayyan)
10	The Five Centuries	Rudolph Glauber
11	The Greater and Lesser Edifyer	Johann Grashoff
12	Chemical Secrets and Experiments	Sir Kenelm Digby
13	The Turba Philosophorum	Arisleus
14	Das Aceton	Christian Becker
15	TBD	These volumes are reserved for the Works of Glauber.
16	TBD	
17	TBD	
18	TBD	
19	TBD	
20	TBD	
21	Alchemical Symbols, Third Edition	Hans W. Nintzel and Philip N. Wheeler
22	The Book of Formulas	John Hazelrigg

23	18 Short Tracts	Hans W. Nintzel
24	Bacstrom's Notebooks, Part 3	Sigismund Bacstrom
25	A Discourse on Fire and Salt	Blaise Vignere
26	The Mineral Work	Johan Hollandus
27	The Vegetable Work	Johan Hollandus
28	Lamspring's Process	Lamspring
29	The Book of Abraham the Jew	Abraham Eleazar
30	Five Short Works of Glauber	Johann Glauber
31	The Metamorphosis of the Planets	Johannes Monte-Snyder
32	Four Works of Roger Bacon	Roger Bacon
33	The Golden Chain of Homer	Homerus, Kirchweger, Nintzel, Wheeler
34	Alchemy Rediscovered and Restored	Archibald Cochren
35	Aurifontina Chymica	John Houpreght
36	The Golden Fleece	Salomon Trismosin
37	The Transmutation of Base Metals into Gold and Silver	David Beuther
38	Sanguis Naturae	Christopher Grummet
39	A Revelation of the Secret Spirit	Giovanni Lambi
40	The Holy Guide, Part 1	John Heydon
41	The Holy Guide, Part 2	John Heydon
42	Secreta Alchymiae	Kalid Persica
43	The Golden Treatise of Hermes	Hermes Trismegistus
44	Potpourri of Alchemy, Part 1	Hans W. Nintzel
45	Potpourri of Alchemy, Part 2	Hans W. Nintzel

www.ingramcontent.com/pod-product-compliance
Lightning Source LLC
Chambersburg PA
CBHW080651190526
45169CB00006B/2074

* 9 7 8 1 5 2 3 7 9 9 6 6 4 *